Scented Containers

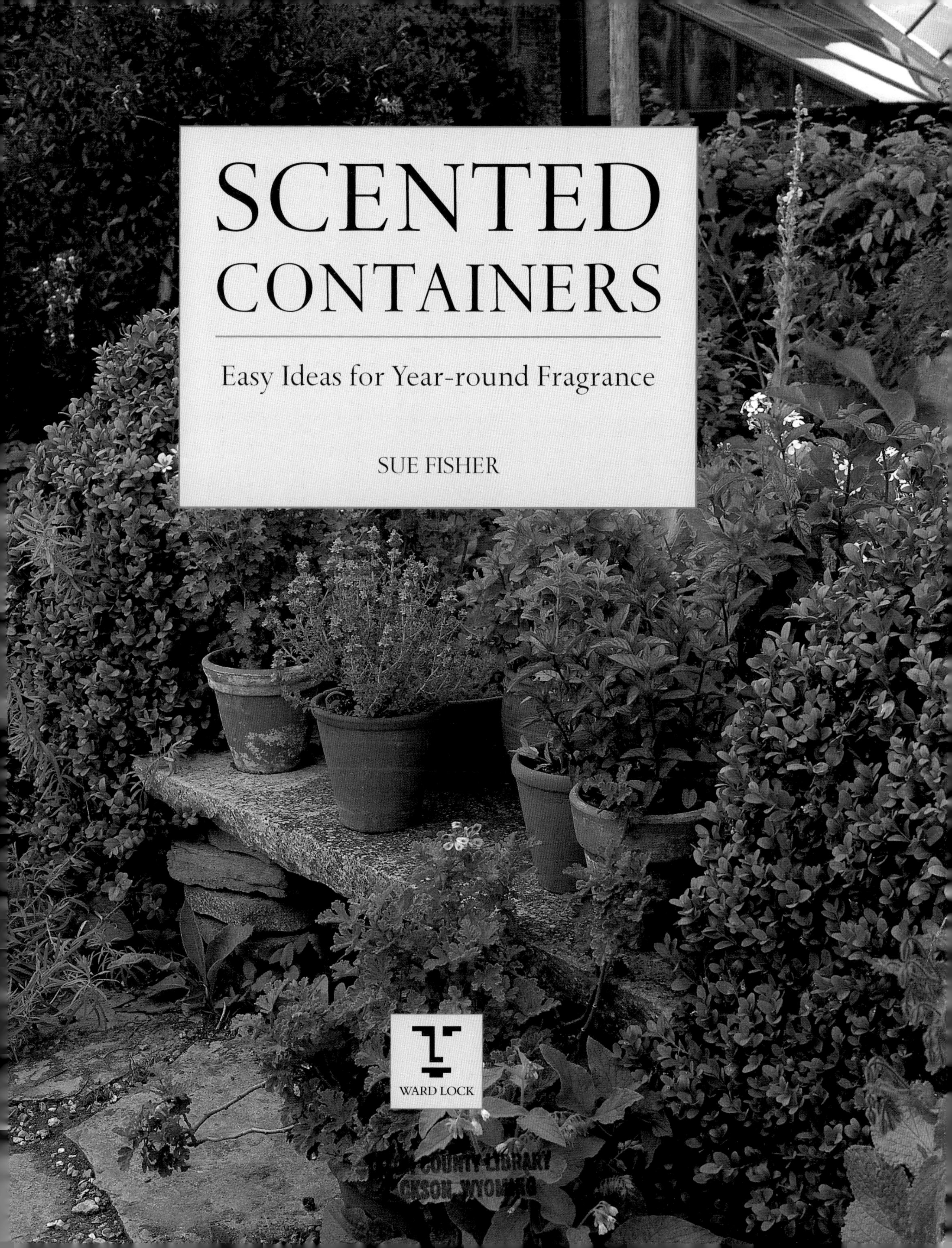
SCENTED
CONTAINERS
Easy Ideas for Year-round Fragrance
SUE FISHER
WARD LOCK

A WARD LOCK BOOK

First published in the UK 1999 by
Ward Lock
Wellington House
125 Strand
London WC2R 0BB
www.cassell.co.uk

A Cassell Imprint
First paperback edition 1999

Distributed in the United States by
Sterling Publishing Co., Inc.
387 Park Avenue South
New York NY 10016–8810

British Library Cataloguing-in-Publication Data
A catalogue record for this book is available from the British Library

ISBN 0-7063-7800-8

Designed by Chris Bell

Printed and bound in Italy by Printer Trento S.r.l.

Contents

Introduction

Our world is dominated by visual images and is rich in technology to the point where sight and sound are vastly predominant. However, in the garden, a place that really should please all the senses, scent has a power and an appeal that are all too often underused or even overlooked completely. Plants with fragrant flowers or aromatic foliage have an enormous attraction, although we rarely acknowledge this dimension to the full. Even though I adore scented plants, it took me some time to realize that they were entirely responsible for the erratic path of my stroll around the garden, as I wandered from a richly perfumed pot of lilies, across the lawn to crush a few leaves of calamint for their pungent, minty aroma, zigzagged back to pick a few stems of sweet peas for the house, before turning to a tucked-away seat to luxuriate in the evening scent of gleaming white tobacco plants.

Portable perfume is best of all, because plants growing in containers can be moved to the exact place where they can be enjoyed to the full. Another huge bonus is that after the flowers have finished or if the plants start to look jaded, they can be moved 'off-stage' and swapped for others that have yet to reach their peak. The sheer beauty of containers lies in

Plants with scented flowers or aromatic foliage can be planted in pots and positioned for maximum enjoyment. *Origanum* 'Kent Beauty' makes a handsome and aromatic table centrepiece.

their tremendous versatility, so no matter how small your patio or garden may be, you can have a constantly changing feast of fragrance and colour. Herbs can provide an edible feast, too, being eminently suited to containers and being aromatic as well as beautiful.

Containers also make it possible to garden in the most awkward of spots, and these are often the places where scented plants can be appreciated to the utmost extent. Paved sites, such as around the main doorway or next to the front gateway, where a waft of scent makes the ideal welcome home after a hectic day, are prime yet neglected positions. For summer, put fragrant flowers by open windows so that their perfume can drift through the house. Around the garden, place scented pots near seats to add a bit of home-grown aromatherapy that will intensify your relaxation. For conservatories and sun-rooms there can be a wonderful seasonal plant-swap, and many shrubs that are exotic yet frost-tender can be treated as 'inside-outside' plants, staying indoors for the winter and moving out onto the patio in summer.

Through the seasons, this book will look at the wealth of wonderful fragrant plants that are available and show how to grow them in all manner of different ways. A host of tips, projects and seasonal reminders will help and encourage gardeners of all levels of experience. There is a plant directory for easy reference, plus all the nitty-gritty practical information on how to plant and care for your containers in order to get the very best results with the minimum of work. So, whether you are new to gardening or are just searching for some fresh ideas, there should be something here to inspire everyone.

Sites such as patios and paths where there is no soil are often where scented plants in containers can be of most benefit. The fragrance of white *Nicotiana* (tobacco plant) is strongest in the evening.

The Nature of Fragrance

WHILE we sit and luxuriate in all the different aromas produced by a multitude of fragrant plants, it is interesting to ponder a little on the whys and wherefores of fragrance. Why, for example, are some plants scented while others are not; what is the purpose of aromatic foliage; and how do we actually smell these plants anyway?

Why Plants are Scented

Scented plants have been valued by gardeners for thousands of years, but plants have these properties for their own benefit, not ours – the fact that we enjoy their fragrant flowers and aromatic foliage is a happy coincidence. However, it is interesting to discover the reasons why plants are scented, and it all centres round survival.

Flowers are fragrant for the purpose of pollination and, therefore, for the continued existence of a plant. In the flower are the plants' reproductive organs, the fertilization of which is dependent on the receipt of

Pelargoniums with scented leaves are splendid foliage plants when grown as individual specimens. The oils contained in the leaves of aromatic plants help to deter pests.

Many lilies have a strong and delicious perfume as well as extremely showy flowers. The large blooms of *Lilium* 'Star Gazer' make a bright and eye-catching display.

pollen brought by insects. Those plants that are fertilized by moths and butterflies generally have the strongest scent because they have to attract insects that fly over considerable distances. By contrast, bees target flowers according to sight rather than smell, which is why bee-pollinated blooms are usually colourful rather than scented. Plants that fertilize themselves also tend to be brightly coloured.

Scent is also a defence mechanism, particularly when it comes to plants with aromatic foliage. In drought-prone regions many plants protect themselves from drying out by giving off an oily vapour from their leaves, which is why so many of the Mediterranean herbs have wonderfully aromatic foliage. The ease with which aromatic leaves release their scent depends on where the essential oil is stored within the leaf.

In many cases, such as with thyme, it is stored near to the surface and therefore released readily when the leaves are warmed by the sun. A few plants, such as bay, store the oil deep within their leaves and so it is only released by crushing and bruising. A number of plants also use scent to ward off attacks by insects or animals by making themselves unpalatable. There are plentiful numbers of such plants, including asphodel, cistus and spurge in Mediterranean regions that are heavily grazed by animals, and extensive colonies of these varieties are a sure sign of severe over-grazing. Garden plants like chrysanthemum and pyrethrum produce odours that repel insects.

However, as the technology of scent analysis improves, a more complex and fascinating picture has begun to appear. When certain plants are eaten by insects, they have been found to emit a chemical to attract other insects that will predate on the attacker. Cabbages that are infested with cabbage white butterfly caterpillars, for example, attract a parasitic wasp that attacks the caterpillar. Other plants that are under attack from insects generate a chemical as a warning to neighbouring plants, and these then produce certain substances to make themselves less palatable. This might even mean that in the future natural plant activators will be used to stimulate resistance, rather than having to rely on harmful pesticides or fungicides.

Opposite: A group of small pots of spring bulbs makes a superb show. Blue hyacinths and *Iris reticulata* 'Harmony' are scented, while yellow *Narcissus* 'Tête-à-tête', *Iris danfordiae* and *I.* 'Katharine Hodgkin' provide a beautiful colour contrast. Early bulb flowers are a life-saving source of nectar for emerging insects such as bees.

The Physiology of Smell

Our sense of smell is thought to be the most primitive of our senses, yet the human body registers a scent in half a second – almost twice as fast as it does pain. Although we no longer need to use smell as an aid to survival as other mammals, birds and insects do, it still has a tremendous effect upon us. A waft of a particular scent can transport us to a place or an incident that happened long ago, and certain scents can affect and even manipulate our moods – hence the fast-growing popularity of aromatherapy.

There is a very close relationship between taste and smell. We sense smell through tiny hair-like rods in the top of the nose, and when we eat and drink, the aroma goes up into the nose so the two senses interact in subtle ways. Perhaps this is why so many scents are described in food-like terms such as lemon, honey and chocolate.

Sensitivity to smell varies from one person to another. Some people can pick up the faintest of fragrances, while others need stronger perfumes to invoke a response. It's worth noting that smoking has a detrimental effect on the sense of smell, sometimes to a considerable degree. By seeking out different fragrances and how they appeal to yourself, it is possible to build up a mental 'library' of preferred scents and to plan your garden for maximum enjoyment.

Planning for Scent

Getting the most from scented plants involves a brief assessment of your garden's layout and how you use it, so that you can site plants where they can be of greatest benefit. The most significant aspect when looking at your garden is shelter, which is a key element in appreciating fragrance to the full. A garden that is open to wind will have its perfumes whisked away, while in a sheltered spot the scents tend to hang around their source or to be wafted a little way on a gentle breeze. The orientation of a site – the amount of sun or shade it receives – is also important, because the vast majority of fragrant plants prefer a reasonable amount of sun. (See page 17 for some shade-tolerant plants.)

Start with the house front and any frequently used gates and pathways. Such areas tend to be viewed as functional and as such do get passed over, but the very frequency of their use should make these sites prime candidates for scented plants. What could be more pleasant than some wafts of perfume *en route* to work or shopping and then when returning wearily back home again? Passers-by appreciate scent too. I get enormous satisfaction from sitting in my living-room, out of sight, and hearing total strangers commenting with delight on the surprisingly strong scent of the *Sarcococca* (Christmas box) in midwinter, or the heavy and delicious evening perfume of *Nicotiana* (tobacco plant) in high summer.

The patio, which is usually adjacent to the house, is the obvious location for a wide variety of containers. The house will provide a boundary to at least one side of the patio and often more, and a garage or next-door house may well provide additional shelter. However, unless

Opposite: Fragrant flowers, particularly powerfully scented ones like this *Nicotiana*, are best mixed with unscented blooms to avoid an overpowering mixture of perfumes.

Right: Plants in small pots can be placed on a low wall or shelf to be nearer the nose. The spring bulb *Muscari armeniacum* (grape hyacinth) has a delicate honey scent.

your patio is very sheltered or is in a courtyard, consider putting up some screens so that it is partly enclosed. Bamboo or split-willow screens look very stylish and are ideal in a formal setting, while hurdles made of woven hazel or willow match a cottage to perfection. Another option is trellis, which can be clothed with climbing plants – variegated ivy is evergreen and excellent for year-round interest, and it can be interwoven with sweet peas for scented summer blooms. Most costly is a brick wall, although its storage-heater effect is wonderful in cold areas for growing plants that are on the borderline of hardiness. When siting a wall or screen, make sure it is placed to shield the seating area from the prevailing wind.

Take full advantage of any low walls that may edge the patio, for their tops are usually just perfect for small pots of plants that are then around nose-level for anyone sitting down. Such a site is ideal for those flowers that shyly protect their beauty unless it is seen at close quarters, such as the delicate markings inside the dwarf *Iris reticulata* or the intricate laced patterns that decorate some *Dianthus* (pinks).

Plants with aromatic foliage are ideally placed near to seats or close to where one will walk, to make it easy absent-mindedly to pick and crush a leaf to breathe in its scent. Herbs are prime candidates for a patio so that they are conveniently close at hand when a few sprigs of foliage are needed in cooking or to decorate a summer drink. Put a pot of rosemary by a barbecue and lay a few twigs on the grill when cooking lamb – or, better still, strip the leaves and use the twigs to make rosemary-flavoured lamb kebabs.

Aromatic herbs in pots can be positioned conveniently close at hand for cooking. Chives and thyme are, a simple yet attractive duo.

However, these are summer delights, and the days on which they can be enjoyed will be in the minority in cold or exposed areas. Here, a conservatory is the ultimate luxury, making it possible to create an absolute feast of fragrance, for this protected environment can house a selection of frost-tender plants with rich and beautiful perfumes. And, once all danger of frost is past in late spring or early summer, plants in containers can be moved out onto the patio to introduce a foreign and exotic touch to the atmosphere. This movement of plants can be a two-way one, for hardy plants – particularly those that bloom in winter – can be grown outside and then brought into the conservatory at their time of glory, to be appreciated at leisure in a sheltered environment.

Consider the siting of plants in relation to windows. On warm days, windows will be open to let scent waft in from outdoors. Climbers can be grown in pots and their stems wreathed around the window, while strongly scented plants in containers can stand underneath. The rich fragrances of jasmine, lilies,

Sites for Scented Plants

1 A fence or hedge will filter wind.
2 A patio is the obvious place for pots.
3 A trellis or split-cane screen provides extra shelter on a patio.
4 The perfume of strongly scented plants will drift indoors.
5 Fragrant climbers around windows will scent the house and patio.
6 Herbs should be handy for the kitchen.
7 A low wall edging a raised bed is ideal for small pots of delicate flowers, which can be brought closer to the nose.
8 An arbour with a seat can be a fragrant retreat.
9 Paving slabs will keep feet dry – place fragrant plants in pots nearby.
10 Trellis, hurdle or cane screening will hide the 'nursery' area.
11 A 'nursery' area can include a greenhouse and space for plants that are growing on or are past their best.

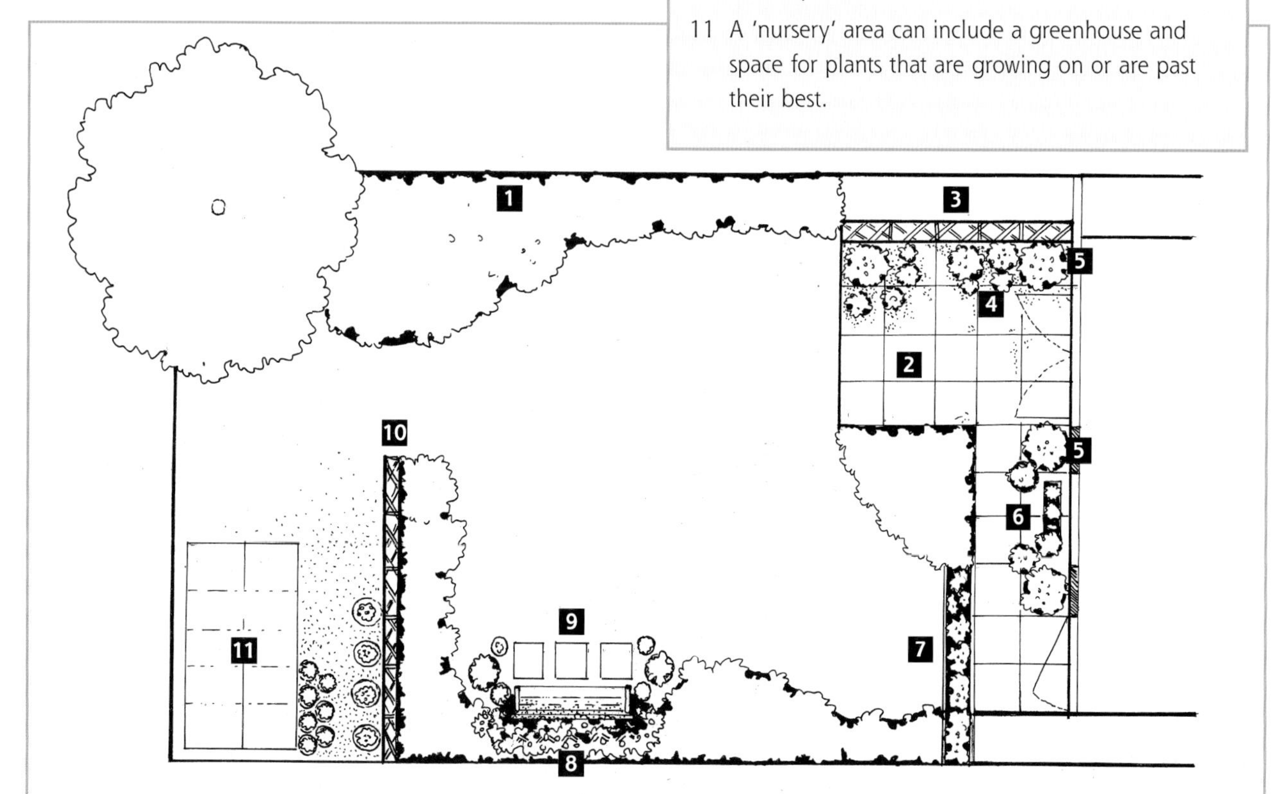

Nicotiana (tobacco plants) and *Matthiola longipetala* (syn. *M. bicornis*; night-scented stocks) can drift in to please the senses even after you have abandoned the garden for the television!

Away from the house, there is something immensely pleasurable in having somewhere to sit apart from the patio – which is, after all, often a little too close to the demands of family and telephone for comfort. Choose a sunny spot and create shelter by putting one of the screens described above on at least two sides of a seat. Place a few paving stones in front, to make the ground warm underfoot as well as to keep your feet dry and comfortable, then gather a collection of your favourite perfumed plants in pots.

Even better than a scented seating corner is a little arbour, which will create the real feeling of an outdoor room. A simple framework can be made using trellis, or you could just buy one of the many different arbours that are available in kit form. Generally this will be a summertime retreat only, so train a couple of summer-blooming climbers over the sides and top – *Trachelospermum* in mild areas, jasmine or honeysuckle where it is cooler, plus sweet peas – and again, a group of pots around your feet.

One word of caution when you are creating a garden using scented plants. Take care not to overdo matters by choosing *just* fragrant plants, because too many perfumes could vie with each other and end up becoming rather nauseating. Mix in a fair percentage of non-fragrant plants to balance things out. When you are selecting unscented plants, it's worth choosing a number of evergreens that have decorative foliage so that your containers will look good right through the year.

Lastly, it's well worth making space in your garden for a tucked-away 'nursery' area. This will be the 'backstage' to your container displays where plants can be discreetly grown on until they are good enough to take centre stage. Once flowering is over, permanent plants can be brought back to grow on for next year, while containers of seasonal varieties can be replanted. If you enjoy propagating your own plants, include a greenhouse, where all sorts of plants can be raised from seed or cuttings, and a cold frame, where they can be hardened off before facing the outside world. For those who have never tried any propagation at all, I urge you to have a go. Raising plants from scratch is enormously rewarding, and I have never quite got over the miracle of a beautiful plant developing from a tiny, wizened seed. Then, when you come to sit and breathe in its wonderful scent, there will be a lovely feeling of proprietary joy too, because it will be all your own work – well, with a bit of help from nature.

Opposite: Shelter is an important consideration to avoid perfumes being whisked away on the breeze. Trellis creates an intimate corner in which to enjoy these scented sweet peas.

Scented Plants for Shade

The following plants will grow in partial shade as well as in sun. Those marked with an asterisk (*) tolerate full shade. All are listed in the Plant Directory on pages 107–123.

Shrubs
Mahonia aquifolium 'Smaragd'
Philadelphus 'Manteau d'Hermine'; *P. microphyllus* (mock orange)
**Sarcococca* (Christmas box, sweet box)
**Skimmia* varieties
Viburnum × bodnantense; V. × juddii

Climbers
Akebia quinata (chocolate vine)
Clematis flammula; C. rehderiana
Lonicera (honeysuckle)

Perennials
**Convallaria majalis* (lily-of-the-valley)
**Glechoma hederacea* 'Variegata' (variegated ground ivy)
Primula veris (cowslip)
**Primula vulgaris* (primrose)
**Viola odorata* (sweet violet)

Annuals and biennials
Asperula orientalis (blue woodruff)
Reseda odorata (mignonette)
Hesperis matronalis (sweet rocket)

Bulbs
Galanthus (snowdrop)
Lilium (lilies, some)
Muscari (grape hyacinth)

Herbs
Petroselinum crispum (parsley)

Choosing Containers and Creating Displays

GARDENERS these days are fortunate in having a vast selection of containers to choose from. Although choice is obviously a very personal matter and your own particular taste will dictate your selection – not to mention your budget, because some containers are on the expensive side – there are a few guidelines that can be useful (see panel opposite). First, though, let us look at the types of pot that are generally on offer.

Cheapest of all tend to be plastic containers, some of which are passable imitations of terracotta or stone. Seasonal summer plants are fine in this type of pot, but the thin sides offer little insulation in winter, so it is best to avoid plastic for permanent plants in cold areas.

Terracotta is very popular and there is a considerable selection of sizes and designs on the market, including those with a glazed, oriental-style finish. Prices vary according to the size and amount of decoration, and also whether the pot is frost-proof or not, an essential consideration in cold areas if the container is to stay outside all year, but not if it is for spring to autumn use only.

Wooden containers come in a whole range of sizes with designs varying from simple windowboxes to elegant Versailles tubs (so called as this style was created at the Palace of Versailles in the seventeenth century). Half-barrels – often recycled from the brewing industry – are large and cheap in comparison to their size, and they are, therefore, excellent for substantial permanent plants such as shrubs and climbers.

Wall pots, half-baskets and wall troughs are flat on one side and can be hung up on hooks or nails. Those made with solid, thick sides – such as terracotta – are very useful for brightening up walls from autumn to spring, as the sides give sufficient insulation to enable hardy plants to be successfully grown. The sheltered environment created by a house wall will often protect slightly tender plants too, which will come through the winter here but are likely to die just a little further out in the garden.

Hanging baskets are immensely useful and mostly inexpensive, apart from a few large and stylish models. I prefer to opt for the cheap and cheerful plastic-coated wire mesh baskets, which, although not at all ornamental, should soon be completely concealed by plant growth if planted properly. This is the most versatile type of basket because the sides and base can be planted as well as the top, to create a globe of glorious colour.

Tall containers such as urns and jardinières are very useful for creating height within a grouping of pots. Those made of stone or wrought iron are the most ornamental – and expensive – although cheaper and reasonably good-looking imitations are available. Old chimneypots are excellent for height too, although genuine ones are becoming increasingly scarce and expensive.

Quite apart from all these different containers that can be purchased from garden centres and other sources, there is a whole host of

A scented-leaf geranium and trailing variegated ground ivy (*Glechoma hederacea* 'Variegata') provide aromatic and attractive foliage in this hanging basket.

Wicker baskets are charming containers for short-lived displays such as this white hyacinth, *Hyacinth orientalis* 'L'Innocence'. Baskets last a long time if varnished and lined with polythene.

other items that can be adapted for use as plant containers, often at little or no cost. Wicker baskets filled with bulbs, annuals or wildflowers look enchantingly pretty in an informal setting and are surprisingly long-lasting if painted thoroughly with clear varnish and lined inside with plastic. I recommend buying good quality handmade baskets rather than cheap imports. Throwaway items such as old sinks and toilets, pieces of land drainpipe, central heating tanks and galvanized metal buckets can all be planted up. Remember to take care that any potential plant container either has drainage holes or the material can be drilled to make some, because plants with waterlogged roots will quickly die. Look around junk yards and auction sales, and keep an eye on any local building work to see if there is anything worth scavenging.

Positioning and Grouping Containers

In places such as a patio where a number of containers will be grouped together, try to choose pots of different sizes so that the contrasting shapes create plenty of visual interest in their own right, rather than relying on the plants alone. However, in the real world it is all too easy to end up with lots of pots of a similar size – particularly if they're offered at a bargain price! In this case, inject more interest by making 'steps' on which to raise several pots to the rear of a group, or even just by turning an

Choosing Containers

- Beware of having lots of different types of pot that will give a jumbled, 'over-busy' impression.
- Stick to one or two materials, and introduce variety by choosing pots of different shapes and sizes.
- Try to visualize which pots will look most in keeping with your house and garden. Stone or stylish terracotta is best in a formal garden, while an informal cottage-style garden could house a selection of 'throwaway' items used as pots.
- Consider buying at least one matching pair of good quality containers – they look very effective flanking a door, gateway or flight of steps.
- Take care to choose frost-proof materials for pots that will be outside all year round in cold areas.
- Think about creating a particular theme, such as glazed pots for an Oriental-style garden, or terracotta for a Mediterranean-style courtyard or patio.

empty pot upside-down to make a base for another. Height can also be added to a group of pots by using tall containers such as chimneypots and jardinières, planting some climbers in containers (see page 52), and with standard plants that are trained to have a couple of feet of clear stem (see page 59).

The obvious site for hanging baskets is suspended from brackets fixed to walls, but other potential sites include fence posts and pergolas. It's also possible to have a free-standing display by using a hanging basket 'tree' made from wrought iron, and these can vary in size to take from two to ten baskets.

When you are siting containers at the front of the house it is an unfortunate fact that security has to be borne in mind. Free-standing pots should be sufficiently large and heavy to be near-impossible to carry off.

Decorate your own Containers

Add an artistic touch to your container garden by decorating or painting plain terracotta, wood or plastic pots. If you possess a motley collection of pots, painting them all in the same colour will create harmony in an otherwise conflicting group. Use an acrylic or oil-based weather-resistant paint for wood or plastic and one of the many coloured stains for wood. For the best overall results, stick to just one mode of decoration for a single area such as a patio. Go for simple and restrained planting in decorated pots to avoid an over-powering and cluttered look.

Prepare containers first by rubbing the surface with medium-grade sandpaper, and then clean the surface thoroughly. Here are a few suggestions.

- Bold vertical stripes in one colour, interspersed with white.
- A stencilled pattern on a plain painted background (either cut out your own design from cardboard or buy ready-made stencils).
- An all-over wash of a single colour, so the original colour of the pot partly shows through.
- Sponging, using a different colour over a base coat.
- An 'antique' look, using a dense sponge to dab paint on unevenly so that some areas are left unpainted.
- Any trellis or wooden supports can be stained to match a painted container.

Ordinary pots are simple to decorate yet look very effective.

Creating a Staged Display

1 *Trachelospermum asiaticum*
2 *Myrtus communis*
3 *Cosmos atrosanguineus*
4 *Mirabilis jalapa*
5 *Nemesia denticulata*
6 *Rosa* 'Norfolk'
7 *Zaluzianskya capensis*
8 *Lavandula angustifolia* 'Hidcote'
9 *Lobularia maritima*

A GROUP of containers looks most effective when it consists of pots with different heights, shapes and sizes that create lots of interest in their own right. For a really attractive grouping, include a couple of large specimens and place several of the pots at the back on supports, so the plants can be seen to full advantage.

The easiest way to create height at the back is to stand pots on other, inverted containers or a similar support, such as a stack of bricks. One or two climbers or wall shrubs can be grown in containers and trained up a wall or fence if available. If not, train them on a tower of canes (see page 49) or on a trellis fan (see page 71).

All the plants listed bloom in summer. Several prefer a sunny sheltered site and will need winter protection in cool areas.

At the back of the arrangement are two evergreens, *Trachelospermum asiaticum*, a climber with creamy-yellow, strongly scented flowers, and *Myrtus communis* (common myrtle), a compact shrub with aromatic leaves and white flowers. The frost-tender perennial *Cosmos atrosanguineus* (chocolate cosmos) has dark red, dahlia-like flowers with a delicious chocolate scent, and another tender plant is *Mirabilis jalapa* (marvel of Peru, four o'clock flower), a tuberous perennial which can also be grown as a half-hardy annual. The clusters of red or yellow tubular flowers open in the afternoon. *Nemesia denticulata* (syn. *N. d.* 'Confetti'), a frost-tender perennial, produces many slender stems topped with clusters of fragrant, pale pink flowers.

Rosa 'Norfolk' (syn. *R.* 'Poulfolk') is a compact rose, slightly wider than it is high, with double yellow flowers that have a good scent. *Zaluzianskya capensis* (night phlox) is a hardy annual with starry white flowers that are strongly scented in the evening.

In the final pot is the evergreen *Lavandula angustifolia* 'Hidcote' (lavender), an aromatic-leaved shrub with dark blue, fragrant flowers, and this is surrounded by *Lobularia maritima* (sweet alyssum), a hardy annual which has white flowers.

Containers placed at different levels create visual interest and allow each plant to be seen at its best. *Lilium regale* (regal lily) is easy to grow and richly perfumed.

Colour Schemes

Attractive colour schemes are as important with container plants as anywhere else in the garden, though it's a lot easier to achieve a harmonious balance of colour with pots that can be moved around to create the desired effect.

First, consider the background when choosing which colours to have. A dark background, such as a brown fence, a dark green hedge or a red-brick wall, is ideal to show up pale flowers and variegated foliage. A lighter background, however, such as a whitewashed wall or pale stone, will make a perfect backdrop to richly coloured flowers and foliage.

The best overall effects are achieved by restricting the number of colours in a group of containers. Just two or three colours, for example, look far more effective than a 'liquorice allsorts' collection of lots of different ones. Another option is to stick to a single colour – such as pink – but use lots of different shades to create variety.

Aromatic foliage plants create a sophisticated colour scheme: scented pelargoniums (front and back), purple sage (back right), mint and thyme (front right). Creeping Jenny (*Lysimachia nummularia*) and *Sempervivum* provide contrast but are unscented.

Wooden half-barrels or large stone troughs are good examples. Hanging baskets can be secured to their brackets with a small padlock and chain, or by wire that is repeatedly wrapped around the bracket. Windowboxes can be held in place with wires going through the drainage holes and round the back of the box to a couple of strategically placed nails.

Apart from these general points, take a good look at your garden to see what other opportunities present themselves for siting containers. A flight of steps can have a series of small, identical pots ranged up one side, while a low wall can have similar pots in a line on top. Wooden shelves on a wall can also house a row of small containers. For a multi-pot plant stand against a wall or fence, consider using a few pieces of wood to make a permanent stage. This way, you can have two or three layers of pots and, because the 'stage' should be completely concealed, it can be made from old timber. Of course, if you wanted to make it into more of a feature, use good quality wood and finish it off with a coat of coloured wood stain. With a bit of imagination and ingenuity, there is hardly anywhere that need be without plants.

Spring – A Fragrant Beginning

After the long, gloomy days of winter, the fragrances of spring are perhaps the most welcome and evocative of all. The first flowers to open have a delicate, almost tentative fragrance, such as that belonging to wild primroses and violets, as they join the snowdrops and crocuses that have already braved the harshness of winter. In the awakening garden there is delight in the first flights of bees and butterflies, emerging sleepily from their winter hibernation and seeking these precious early blooms for much-needed nourishment. As the days become warmer and lengthen, many more flowers join the spring pageant – colourful wallflowers and richly scented stocks; hyacinths, the clustered blooms of narcissi and lily-of-the-valley – and shrubs such as pieris and skimmia open their long-dormant buds into bloom.

Opposite: Skimmias offer exceptionally long-lasting interest. *Skimmia japonica* 'Rubella' holds showy red flower buds all winter which finally open to clusters of fragrant white flowers in spring. *S. j.* subsp. *reevesiana* bears fewer flowers but has berries from autumn to spring.

Below: Bulbs are synonymous with spring and some are very sweetly scented, including the double daffodil *Narcissus* 'Cheerfulness'.

Bulbs – A Pageant of Spring Colour

Bulbs are extremely versatile, being quick and easy to grow, immensely colourful and ideal for a whole range of containers. There is an excellent range of scented bulbs to choose from, and their flowering times vary considerably too, so by choosing a number of different varieties it is possible to have bulbs in bloom from the middle of winter right up to the arrival of summer. Bulbs should have been planted the previous autumn, but don't despair if you've missed out, as most garden centres and nurseries offer pots of ready-grown bulbs in spring that are just perfect for popping into containers.

Narcissi, which bloom in early to mid-spring, are nearly all scented to some degree, but it is the jonquil, tazetta and some of the double types that have the richest and most powerful perfume. All are excellent in pots, growing to between 15 and 45cm (6–18in) high and bearing small, cupped flowers, sometimes singly but usually in small clusters. By carefully choosing a range of varieties that flower at different times, it's possible to have narcissi blooming over a long period through the spring. The double types are mostly early flowering; tazetta narcissi flower in early to mid-spring; while jonquils bloom in mid- to late spring. A sunny, sheltered site gives the best results, and for good performance in future years, put the pots in a sun-baked spot for the summer and move under cover to a frost-free place in winter.

The double daffodils, some of which will be familiar from being sold as cut flowers, tend to be first to flower. Indoors, their strong scent can occasionally be almost overpowering, but outside the fragrance is reduced to a pleasant, rich perfume. *Narcissus* 'Cheerfulness', which is creamy-white, goes well with the golden flowers of *N.* 'Yellow Cheerfulness', while *N.* 'Sir Winston Churchill' is a top-performing variety that is white with orange florets. All are multi-headed, with three or four flowers to each stem.

A Group of Pots for Spring

1 *Akebia quinata*
2 *Syringa meyeri* var. *spontanea* 'Palibin'
3 *Daphne odora* f. *alba*
4 *Pieris japonica* 'Little Heath'
5 *Hesperis matronalis*
6 *Melissa officinalis* 'Aurea'
7 *Hyacinthus orientalis* 'Blue Giant'
8 *Narcissus* 'Pencrebar'
9 *Narcissus* 'Flower Drift'

BULBS are synonymous with spring, and there are many delightful scented varieties to choose from, particularly among the different hyacinths and narcissi. However, don't overlook other superb plants, including *Akebia quinata* (chocolate vine), a climber with small, red-purple, chocolate-scented flowers and attractive five-lobed leaves. *Syringa meyeri* var. *spontanea* 'Palibin' (syn. *S. palibiniana*), a dwarf variety of lilac, bears large clusters of fragrant, lavender-pink flowers. *Daphne odora* f. *alba* (syn. *D. o.* var. *leucantha*) is a dome-shaped evergreen shrub, which produces small clusters of white, very fragrant flowers over a long period. It needs winter protection in all but mild areas. *Pieris japonica* 'Little Heath' is a compact evergreen shrub, with green and white variegated leaves and racemes of white, urn-shaped, honey-scented flowers.

As spring moves towards summer, the easily grown biennial *Hesperis matronalis* (sweet rocket, dame's violet), with its mauve or white flowers, brings fragrance to the evenings. Untidy in habit, it is best tucked away behind the other plants. *Melissa officinalis* 'Aurea' (lemon balm) has yellow and gold leaves that form a neat clump and smell strongly of lemon when crushed.

The bulbs include *Hyacinthus orientalis* 'Blue Giant', which has large flowers with a powerful scent, and two daffodils, *Narcissus* 'Pencrebar', which is a double-flowered variety with blooms like little golden roses, and *N.* 'Flower Drift', which has white outer petals and a red centre.

Tazetta narcissi, which also bear flowers in bunches on each stem, do extremely well in pots. There are several lovely hybrids, including *N.* 'Geranium', which is white with an orange cup; the dwarf *N.* 'Minnow', which is creamy-white with a lemon cup; and *N.* 'Grand Soleil d'Or', a showy, golden-yellow variety with orange cups. One to avoid is *N.* 'Paper White', a popular variety for indoor forcing that has a variable performance outside. See page 83 for jonquil narcissi.

A colourful spring group with skimmias and blue grape hyacinths (*Muscari armeniacum*). Red *Tulipa praestans* 'Fusilier' contrasts with attractive conifer foliage.

Growing Bulbs for Indoor Flowers

Prepared hyacinths and narcissi such as *N.* 'Paper White' and *N.* 'Cragford' are widely sold to 'force' for early flowers in the house, but it is not widely realized that most bulbs can be grown to bloom indoors too, usually flowering a little earlier than they would do outside.

Choose varieties that are naturally short and plant in pots of bulb fibre or potting compost so that just the 'nose' of the bulb is showing. Water the compost so it is moist but not waterlogged. Place the pots in a cool, completely dark, well-ventilated place. Check regularly, water if the compost is drying out and bring into a cool, light room when the shoots are about 5cm (2in) high. Once the leaves have greened up, the bulbs will tolerate a warmer room, but preferably with a temperature no higher than 15°C (59°F), and out of direct sunlight.

After flowering, remove the dead flowerheads, and continue feeding and watering until the leaves have yellowed. Bulbs that have been forced should not be treated in the same way again, but they can be planted out in the garden where they should continue to bloom in future years.

Opposite: Blue hyacinths have a beautiful perfume and make a lovely contrast to yellow daffodils. The subtle shades of *Euphorbia myrsinites* (right) and *Veronica peduncularis* 'Georgia Blue' (left) help to pull this grouping together.

Hyacinths, the showiest of all the scented bulbs, have massive heads of richly scented flowers, which can become so top-heavy that a little discreet staking with canes and string may be called for. There is a large range of colours to choose from, and although blue tends to be most popular there are also red, pink, white, apricot and yellow varieties. When buying hyacinths in autumn, take care not to confuse those for growing outdoors with bulbs that have been specially treated for forcing and flowering indoors in winter. To avoid the common problem of flower size deteriorating in future years, be sure to plant deep enough with at least 10cm (4in) and preferably 15cm (6in) of soil over the top of the bulb. Another useful tip for good future performance is to deadhead hyacinths by removing the florets only – not the whole flowerhead – because the flower stem actually functions as a leaf. Hyacinths are susceptible to frost damage, so in cold areas they are best grown in an unheated greenhouse or cold frame for the winter. By flowering time, the worst of the frosts are usually past and the containers can go outside.

Tulips are generally thought of as being all show and no scent but a few varieties are fragrant. They bear flowers on tall stems, and as they are formal in appearance are ideal near the house. There are the early singles *Tulipa* 'Generaal de Wet' (bright orange) and *T.* 'Bellona' (golden-yellow), and two varieties of the exotic parrot tulips, *T.* 'Black Parrot' (deep purple) and 'Orange Favourite' (orange-scarlet), with showy, fringed flowers.

Successful Bulb Growing

- Plant at the right time. Narcissi and early-flowering bulbs, such as crocus, need a long growing season and should be planted in late summer to early autumn. Plant hyacinths in mid-autumn, and tulips in late autumn. Snowdrops are best planted 'in the green', immediately after they have flowered.

- Soggy compost can cause bulbs to rot and die. Pots must have drainage holes (some decorative bulb pots don't). Ensure good drainage by putting stones or crocks in the base of each pot and raising it just off the ground, using pot feet or something similar, so that surplus water can drain away.

- In winter, stand bulb pots in a sheltered site outside or in a cold frame to avoid danger of waterlogging and frost damage.

- Mixing perlite into the compost – using about one part of perlite by volume to four parts of compost – helps to alleviate the problems of cold and waterlogging. Perlite acts as an insulator and also absorbs surplus water.

- For a good performance next year, let the leaves die back naturally because this is how the bulb builds up energy for the future. After flowering has finished, remove the dead heads and feed with a high-potash fertilizer. Continue to water regularly until the leaves have yellowed.

- If you want to keep your bulbs for next year but to use the containers for summer plants, dig a trench in an out-of-the-way corner of the garden and transfer the bulbs there to die back, placing them at the same depth as they were growing previously. To be sure of finding the dormant bulbs when planting time comes round again in autumn, lay a strip of wire netting in the base of the trench first and clearly label each variety.

A Cone of Spring Bulbs

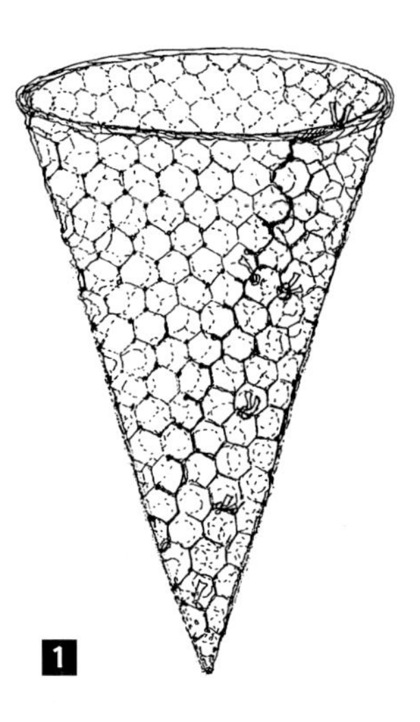

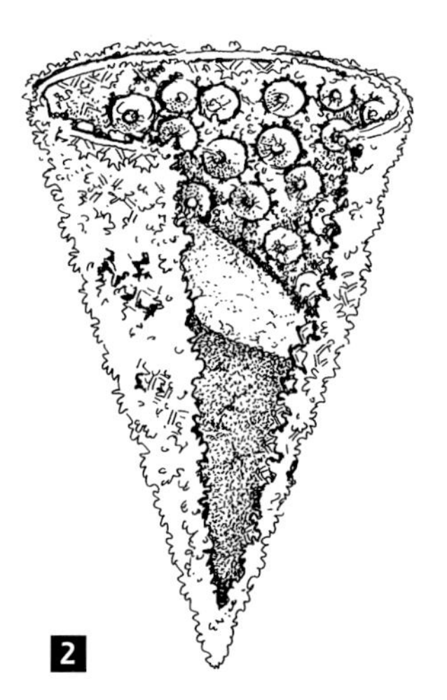

WIRE netting cones planted with dwarf bulbs make a supremely attractive and unusual display that will create a real talking point. Suitable bulbs to use for this include *Crocus chrysanthus* 'Cream Beauty, C. c. 'E.A. Bowles' or C. c. 'Snow Bunting', *Muscari armeniacum* and the most compact varieties of *Narcissus.*

Take a piece of ordinary wire netting or chicken wire – the mesh size should be 2.5cm (1in) – and form it into a cone shape. The size of the piece can vary according to the eventual desired size of the cone and how many bulbs you want to put in it. Tie the overlapping edges together.

Turn the cone upside down and line the bottom half with moss; you can use either sphagnum moss or an imitation moss made from coir. Then press the bulbs into the moss, making sure the tops are pointing outwards. Place the bulbs close together so that you have a good display. Fill the remaining space with soil-less compost and water well, using a watering can fitted with a fine rose. Repeat the procedure for the top half, but be sure to leave enough room for the netting to be folded over to enclose the bulbs at what will be the base of the cone when it is displayed.

Put the cone in a cool, dark place, such as a garage or garden shed, wrapping it in black polythene if necessary to make sure it is completely dark. Keep the cone moist and check regularly on the state of growth. When shoots begin to appear, bring the cone into a light, cool but frost-free place and take care to water more frequently as growth increases.

Stand the cone on a pot filled with compost for display. It can go outside but should be in a sheltered spot because the bulbs are close to the surface and will be vulnerable to frost damage. These cones look extremely attractive if grown in pairs, placed to flank a door, gate or flight of steps.

1 Take a piece of wire netting and form a cone, tying the edges together with wire. Line the bottom half with moss.
2 Press bulbs into the moss with the tops facing outwards and making sure they are placed close together. Fill with compost and repeat for the top half. Leave enough wire netting to fold over to enclose the base. Place in a cool, dark but frost-free place.
3 When shoots appear, move the cone into a light, cool place and stand it on a pot full of compost for display.

Tulips flower in mid- to late spring, as does *Muscari armeniacum* (grape hyacinth), which has cone-shaped blooms of deep blue that have a honey scent.

The ways in which bulbs can be grown are many and varied. Easiest of all is to grow different varieties in individual pots, so they can be moved around to find a pleasing combination and also brought on and off show according to the flowering time. Bulbs go well with other spring bedding plants, such as wallflowers and forget-me-nots, so a larger, handsome container could be planted with a combination of two or three different plants as well as bulbs. Where space is really limited, it's great fun to try 'layering' several different bulbs in one large pot with, for example, tulips as the bottom layer, narcissi in the middle and crocus on top. Amazingly, the lower bulbs manage to find their way round the top ones so the whole container is absolutely packed with colour (see page 73).

Seasonal Spectaculars – Spring Bedding Plants

For masses of cheap and cheerful colour in mid- to late spring, there's nothing to beat spring bedding plants. These are biennials, which are grown from seed sown the previous summer, transplanted in autumn and discarded after flowering. Plants can be bought in autumn, although if you have a spare patch of ground it is easy and much cheaper to grow your own from scratch. Sow outside in a nursery bed in late spring or early summer, gradually thin seedlings to 15cm (6in) apart and transplant them into containers in early autumn.

Most colourful and varied are the gorgeous blooms of that old cottage-garden favourite, *Erysimum cheiri* (wallflower). Of the selection of varieties on offer, the dwarf forms are particularly good for containers, either for growing on their own or for planting under taller bulbs such as tulips. Ready-grown plants are often sold as mixed colours, although an increasing number of sources now offer individual shades. However, if you want a particular colour it may be necessary to grow your own from seed, choosing varieties from the Bedder Series, which offers flowers in pale yellow, golden-yellow, orange or scarlet.

Separating Stocks

Stocks (*Matthiola* spp.) are showy, colourful plants with strongly fragrant flowers that provide first-rate spring and summer colour. However, there are not only different annual and biennial species but also several different strains of stocks, which causes understandable confusion over what to sow and when.

Biennial stocks
Sow biennial stocks outside in a seedbed or in pots in midsummer. In cold areas, overwinter them under cloches or in an unheated greenhouse and plant out in early spring. The group includes Brompton Series, East Lothian Series and the Legacy Series.

Annual stocks
Sow annual stocks in late winter under glass at 10–18°C (50–64.5°F) to flower in late spring to early summer. They can also be sown outside in late spring. The group includes 'Cinderella', Midget Series and Ten-week Series.

To cause even more confusion there are two hardy annuals. *Matthiola longipetala* (syn. *M. bicornis*; night-scented stock) is totally different in appearance to all the showy varieties listed above. It has a sprawling habit and bears small, starry, white to mauve flowers with a powerful clove scent. *Malcolmia maritima* (Virginian stock) is a completely different species but is still sweetly scented and easy to grow. It forms an upright to spreading plant with spikes of white, pink, red or purple flowers. Sow outside in mid- to late spring.

Dianthus barbatus (sweet william) is another popular cottage-garden flower, which has heads of colourful flowers borne in late spring and early summer. Some varieties have flowers of a single colour only, such as red, pink, purple or crimson, while the bicolour forms have an attractive 'laced' appearance similar to that of auricula primulas – and indeed there is one variety of sweet william that is called *D. b.* 'Giant Auricula-eyed'.

Stocks are magnificent, producing columnar heads of richly scented blooms in colours that include red, rose-pink, lavender-blue, dark blue and white (see panel above).

Last but certainly not least is *Hesperis matronalis* (sweet rocket, dame's violet), another old favourite that, unlike the plants mentioned above, is certainly not neat or compact in habit, although in my view it is an absolute 'must-have' for the sweet evening scent of its mauve or white blooms. Growing to around 0.9–1.2m (3–4ft), with tall stems that have a regrettable tendency to flop over, this is a definite candidate for the back of a group of containers. Because the roots will be hidden, don't bother planting sweet rocket in decorative containers but save any black plastic shrub pots and use those instead.

Native Beauty with Wildflowers

The beauty of woodland spangled with primroses and violets or a field full of golden cowslips is the very epitome of spring, although such sights are sadly becoming rarer as so-called progress destroys the countryside.

Planting a Wildflower Trough

A CONTAINER of wildflowers makes a wonderful spring display that is unrivalled for charm and prettiness. Choose a simple trough to match the simplicity of the flowers, such as one made from plain wood or terracotta. Wildflowers also look marvellous in wicker baskets (see page 19).

Put a good layer of crocks in the base to aid drainage, then part fill the trough with a soil-based potting compost. For a trough that is 45cm (18in) long you will need: three *Primula veris* (cowslip), either well-grown clumps lifted from the garden or ready-grown plants in 13cm (5in) pots; five small plants of *Glechoma hederacea* 'Variegata' (variegated ground ivy); and four *Viola odorata* (sweet violet). When all the plants are in position, fill the gaps with compost and water thoroughly.

The violets will flower first, joined later by the cowslips, to make an attractive contrast of blue and yellow. Variegated ground ivy will trail its long stems of rounded, grey-green and white leaves over the sides to soften the straight lines of the trough, and the leaves give off a pungent, refreshing scent when they are crushed.

1 *Primula veris*
2 *Glechoma hederacea* 'Variegata'
3 *Viola odorata*

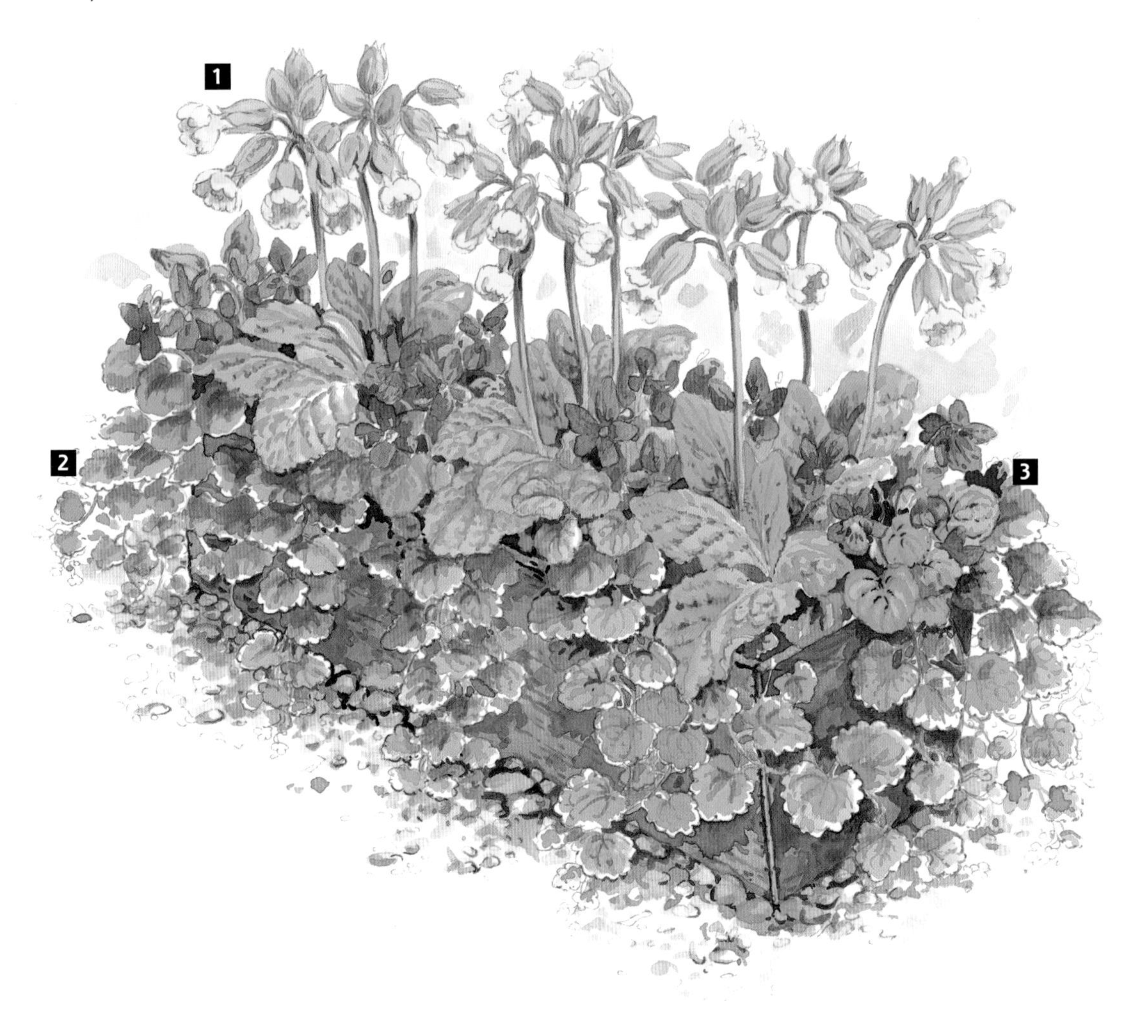

Planting a Large Container for Year-round Interest

A SUBSTANTIAL container such as a large terracotta pot or a wooden half-barrel is a first-rate investment, because it can be creatively planted to give interest right through the year. For the centrepiece choose a good-sized evergreen, which will remain in place all year round. *Choisya* 'Aztec Pearl' (Mexican orange blossom) is suitable for a sunny, sheltered site. Around this central plant put in seasonal flowers, which will be changed at least twice a year and whose flower colours will complement or contrast with the shrub.

For spring to early summer blooms plant the container in autumn with biennials and spring bulbs; the planting illustrated has five *Erysimum cheiri* 'Yellow Bedder'

Spring to early summer

Summer to autumn

(wallflower) and ten blue hyacinths, such as *Hyacinthus orientalis* 'Queen of the Blues'. In late spring to early summer replace these with summer-flowering plants, such as annuals, tender perennials and lilies; the planting illustrated includes three *Nemesia denticulata* (syn. *N. d.* 'Confetti') interplanted with three groups of five bush *Lathyrus odoratus* (sweet pea) and five *Lilium regale* (regal lily). At every replanting replace some of the old soil with fresh potting compost and add some controlled-release fertilizer in spring.

An alternative planting for a sunny spot might include the golden-leaved *Choisya ternata* 'Sundance' as the centrepiece, with, for spring, orange or red wallflowers and blue *Muscari* (grape hyacinth). The summer planting could include *Cosmos atrosanguineus* (chocolate cosmos), white lilies, pale pink *Dianthus* 'Doris' (pink) and *Lobularia maritima* (sweet alyssum).

In a site that is partly shaded or more exposed a good central plant to choose is *Mahonia aquifolium* 'Smaragd', which has glossy green leaves and panicles of yellow spring flowers. For spring surround the mahonia with double-flowered narcissus and *Muscari* (grape hyacinth), replacing these for summer with lilies as before, *Nicotiana* (tobacco plant) and *Asperula orientalis* (blue woodruff).

1 *Choisya* 'Aztec Pearl'

Spring to early summer

2 *Erysimum cheiri* 'Yellow Bedder'

3 *Hyacinthus orientalis* 'Queen of the Blues'

Summer to autumn

4 *Nemesia denticulata*

5 *Lathyrus odoratus*

6 *Lilium regale*

But some of these wildflowers are superb in containers, where they are a joy to look at as well as benefiting bees and butterflies. *Primula vulgaris* (primrose) and *Viola odorata* (sweet violet) do well in shade for early spring, while *Convallaria majalis* (lily-of-the-valley) bears its small stems of exquisitely scented white flowers in late spring. *Primula veris* (cowslip) prefers sun and looks gorgeous with blue *Myosotis* (forget-me-not). I find that when they are grown in good garden conditions, these plants perform way above their wild standards, with a single cowslip, for example, producing up to 15 stems topped with golden, sweetly scented blooms. All these plants have the advantages that they are not only perennial but they often self-seed readily to provide lots of plants for future years.

Permanent Plants – Shrubs and Climbers

Long-lived plants can make a tremendous contribution to the container garden by providing a permanent framework of foliage that is greatly appreciated in winter and early spring. Although such plants are all too often overlooked in favour of seasonal flowers, bear in mind that a combination of different plants will give the best results, with shrubs providing an excellent backdrop for short-lived flowering plants as well as being handsome in their own right.

Just a handful of shrubs provides fragrant spring blooms. A dwarf form of lilac, *Syringa meyeri* var. *spontanea* 'Palibin', bears dense heads of fragrant, lavender-pink flowers, while the most compact of the spring-flowering viburnums is *Viburnum × juddii*, which has sumptuous white, pink-flushed flowerheads with a strong clove scent. Both of these shrubs are spectacular when in flower but are comparatively dull for the rest of the year. They need a large container – a wooden half-barrel is ideal – and this allows for seasonal plants to be grown under, around and even through the shrub to disguise the lack of interest when it is not in flower. Sweet peas are ideal to grow through these shrubs to spangle them with scented summer blooms, while woodruff – either the blue-flowered *Asperula orientalis* or white *Galium odoratum* – could create a carpet beneath the shrubs.

The chocolate fragrance of *Akebia quinata* is just one reason why this climber deserves to be more widely grown.

Pieris, however, are evergreen and so have year-round appeal, as well as being more compact in habit. In spring they bear short stems of urn-shaped, honey-scented flowers that are a magnet for bees, and they produce bright red young growths that look very handsome but that are susceptible to frost damage. Grow pieris in a sheltered spot and keep some horticultural fleece handy for a night-time cover-up if a late frost threatens. The forms with variegated foliage, such as *Pieris japonica* 'Little Heath' and *P. j.* 'Variegata', provide extra interest all year round. Alternatively, choose compact forms such as *P. j.* 'Debutante', *P. j.* 'Dorothy Wyckoff' and *P. j.* 'Purity'. There are also a number of other larger varieties of pieris that will be fine growing in pots for a couple of years, but which will need to be planted out in the garden borders in the longer term.

Daphne odora is another excellent evergreen, forming a rounded, dome-shaped bush with pinky-white flowers that are an absolute delight, lasting for many weeks and giving off a strong lily-of-the-valley scent. Of the varieties available, *D. o.* 'Aureomarginata' is the hardiest, but to avoid frost damage to both roots and foliage it should still be moved into a greenhouse or conservatory for the winter in all but the mildest areas. An unheated structure should provide adequate protection.

Akebia quinata is another delightful and unusual climber that benefits from winter protection in cold areas. Many small, red-purple, chocolate-scented flowers are produced among leaves that are also attractive, being fresh green in colour and five-lobed. A vigorous plant, it needs a large tripod or a similar support up which to scramble.

Buying Frost-tender Seasonal Plants

Garden centres generally begin to stock frost-tender annuals and tender perennial plants very early in the year. In cold areas, containers can be planted up under cover in a greenhouse or conservatory and grown on before being moved outside in early summer, but do take care not to put plants outside too early to be killed by a late frost. It's far better to let the nurseries take the risk and to delay planting your containers until late spring when the danger of frost is past.

Until recently the only choice was between growing your own plants from seed or buying large, expensive, ready-grown ones. Now there are two good options. The first is ready-grown seedlings, which come in one pot and are ready to prick out. The second choice is young plantlets – sometimes called 'baby plants', 'tots' or 'plugs' – growing individually in tiny pots, which can be potted into 8cm (3in) pots and grown on in the greenhouse or on a warm windowsill. In some cases it's possible to take cuttings from the new shoots in early spring, so you can get even more plants for your money.

All frost-tender plants raised under cover need to be hardened off – that is, acclimatized gradually to outside conditions. This is best done over a period of a couple of weeks. Start by putting the plants outside for short periods during the day and bringing them back in at night, then gradually extend the amount of time they spend outside.

A 'Wedding Cake' Layer of Pots

A TIERED display of three pots, one on top of the other, will make a real talking point. Take three matching containers, the largest at least 60cm (24in) across and the smallest about 15cm (6in) across. Glazed containers are often sold in sets of three or five, and these make an excellent and reasonably priced option.

Prepare each container for planting by placing a layer of crocks or drainage material in the base of each pot. The drainage layer should be about 8cm (3in) deep. Fill all the pots almost to the top with compost, then stand them one on top of the other, with the bases of the top two sunk a short way into the compost of the ones beneath for stability. You are now ready to plant.

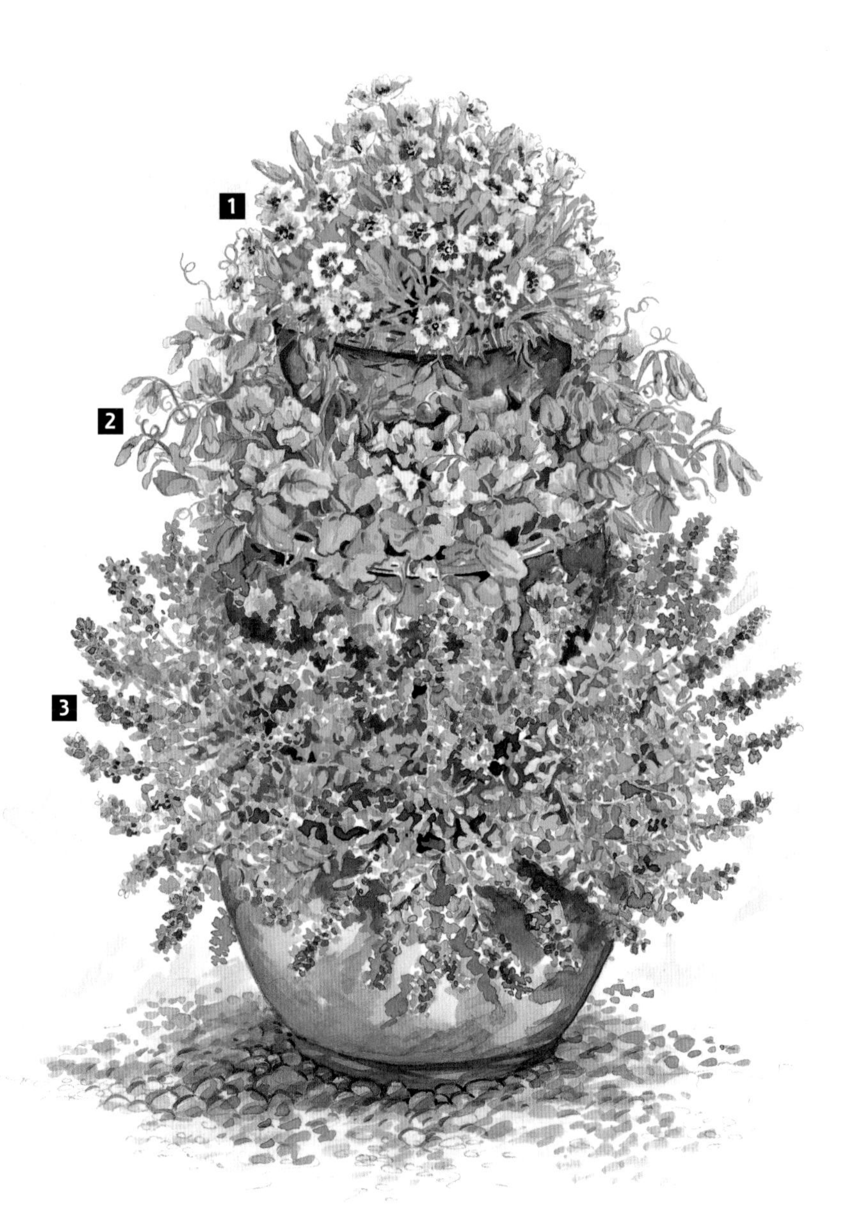

For the topmost pot choose a small plant such as a *Dianthus* (pink); some varieties have flowers that are delicately laced or patterned and are best seen at close quarters. In the centre put several groups of a bush *Lathyrus odoratus* (sweet pea), such as *L. o.* 'Cupid's Pink' or *L. o.* 'Snoopea'. In the lowest pot put five plants of *Nepeta* × *faassenii* (catmint), using small plants grown in 10cm (4in) pots. The sweet peas are annuals and will need to be replaced every year, but the other two are perennials. Stand this display in a sunny place, preferably in a corner or an out-of-the-way spot to avoid any danger of its being knocked over.

Herbs would be an excellent alternative in this type of display. For a sunny site plant *Mentha pulegium* (pennyroyal) in the top pot, *Ocimum basilicum* 'Purple Ruffles' (purple basil) in the centre, with *Origanum vulgare* 'Aureum' (golden marjoram) and *Thymus serpyllum* 'Pink Drift' (thyme) in the bottom pot.

1 *Dianthus*
2 *Lathyrus odoratus*
3 *Nepeta* × *faassenii*

Seasonal Reminders

Propagation

- Snowdrops that have formed large clumps can be divided in early spring.
- Seeds of many annual flowers can be sown now. Half-hardy annuals need protection from frost and are best grown under cover with a little heat, while hardy annuals can be sown outside. Half-hardy annuals can also be sown outside in late spring. Towards the end of spring, harden off plants raised under cover by standing them outside for increasing periods of time over a couple of weeks, then plant them out when all danger of frost is past.
- Sow biennials in a nursery bed in late spring.
- Take cuttings from the new shoots of overwintered tender perennials such as pelargoniums and nemesia.
- Hardy herbaceous perennials can be divided in early spring.
- *Lathyrus odoratus* (sweet pea) can be sown under cover in early spring. Sow outside from mid-spring. Plants sown in autumn should have their tops pinched out when they are 10cm (4in) high.
- Sow annual herbs such as *Ocimum basilicum* (basil) and *Anethum graveolens* (dill). Make several successive sowings of *Petroselinum crispum* (parsley) for a continuous supply. Many perennial herbs can also be grown from seed sown now. Some herbs, such as *Origanum* spp. (marjoram), *Satureja montana* (winter savory) and *Thymus* spp. (thyme), can be propagated by cuttings or layering.
- Divide established clumps of *Allium schoenoprasum* (chives).

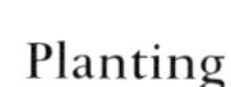

Tender perennials offer exceptional value for money as they last for years if over-wintered in a frost-free place. Dark red *Cosmos atrosanguineus* (chocolate cosmos) has a delicious fragrance.

Planting

- Plant lily bulbs in early spring in deep pots, with several bulbs of the same variety planted together for best effect. Protect the emerging shoots from slugs and snails.
- Plant prepared freesia corms in pots under cover, ready for summer flowering.
- In early spring, pot up overwintered tubers of *Cosmos atrosanguineus* (chocolate cosmos) and *Mirabilis jalapa* (marvel of Peru) into moist compost. Water sparingly until growth is well under way.

Pruning

- Prune roses in early spring.
- Most evergreen shrubs, such as choisya, mahonia and sarcococca, can be pruned if overgrown.
- Hard prune *Clematis flammula* and *C. rehderiana.*
- Trim shrubby herbs such as *Laurus nobilis* (bay), *Rosmarinus officinalis* (rosemary) and *Thymus* spp. (thyme).

General

- Top-dress containers of perennial plants (see page 100).
- Water pots more frequently as the days grow warmer and lengthen.
- Overwintered plants that are on the borderline of hardiness can now be moved outside. Tender plants that require a totally frost-free environment should stay inside until early summer.

Summer – Plentiful Perfumes

Summer is a time of glory, with wonderfully long, warm days and the garden full of delicious heady scents. Seasonal flowers, such as sweet peas, tobacco plants, stocks and mignonette, will be starting to bloom now, and the vast majority should continue to do so right through the summer. Some frost-tender perennials smell almost good enough to eat, particularly *Cosmos atrosanguineus* (chocolate cosmos) and *Heliotropium* (heliotrope, cherry pie). Lusciously perfumed lilies bring an exotic touch to the garden, which can be further intensified by moving any tender shrubs from the greenhouse or conservatory out onto the patio.

On a more homely level, pinks, mock orange, lavender, jasmine and roses supply a multitude of familiar and much-loved fragrances. Add the many herbs and other plants with aromatic foliage, and the gardener is in a sensual heaven.

Summer Spectaculars – Annuals and Tender Perennials

The short-lived annuals and tender perennials offer exceptional value for money in terms of the amount of blooms produced and the sheer longevity of flowering. Given good potting compost and regular watering and feeding, all will bloom for months and many will continue to do so until the first frosts.

Opposite: Handsome in both flower and foliage, *Heliotropium* is also a favourite with butterflies.

Right: Much loved for the captivating fragrance of its flowers and foliage, lavender comes in many forms including *Lavandula stoechas* subsp. *pedunculata* (French lavender) which has unusually shaped flowers. However, this species is more frost-tender than many other varieties of lavender.

Evening-scented plants such as *Nicotiana* are summer essentials. The two scented-leaf pelargoniums have delightful aromatic foliage.

Frequent deadheading, although fiddly, ensures a long flowering period, because plants that are allowed to set seed may feel that their job is done. Sweet peas are excellent examples of this and will stop blooming if not deadheaded regularly.

Scents and appearances in annuals vary enormously. Some plants, such as *Matthiola incana* (stocks), *Nicotiana* (tobacco plant) and *Centaurea moschata* (syn. *Amberboa moschata*; sweet sultan), have substantial, showy and powerfully scented blooms on tall stems and will make a good display in their own right. Others are small, with a more subtle fragrance, and are ideal for infilling – *Lobularia maritima* (sweet alyssum) and *Asperula orientalis* (blue woodruff), for example. Blue woodruff, unusually for an annual, also does well in shade which makes it doubly useful.

A couple of annuals are no great shakes to look at, but any deficiencies in this department are more than compensated for by their strong scent. *Matthiola longipetala* (syn. *M. bicornis*; night-scented stock) is a prime example, with tiny, star-shaped flowers borne on a plant that is, in truth, rather untidy in appearance, but the rich clove scent has a strength that is out of all proportion to the flower size. *Reseda odorata* (mignonette) is taller and untidy with a wonderful evening fragrance. Lastly, don't forget annual climbers, which are tremendously useful for infilling as well as for creating a handsome potful in their own right, particularly *Lathyrus odoratus* (sweet pea), a favourite without which no scented garden would be complete (see page 49).

Annuals are raised from seed, and the ease with which they can be grown depends on whether the variety is a hardy or half-hardy annual. The difference is important. Hardy annuals tolerate frost and can be sown outside, in either autumn or spring, or under cover for quicker results. Half-hardy annuals won't tolerate frost and need to be sown under cover in spring in cold areas.

However, if you're not keen on growing your own, there are plenty of nurseries and garden centres offering ready-grown plants. Just take care not to buy and plant out until all danger of frost is past – let the nursery take the risks. A good compromise is to use ready-germinated seedlings and 'starter' plants.

Most fragrant annuals prefer sun but *Asperula orientalis* (blue woodruff) thrives in shade and is immensely useful for growing underneath a larger plant to cover the soil.

A Group of Pots for Summer

THERE need be no shortage of perfume during the summer months. Clothe walls, fences and other supports with climbers to bring fragrant flowers near the nose and where the scent can drift indoors. The honeysuckle *Lonicera periclymenum* 'Belgica' (early Dutch woodbine) is a twining climber with sweetly scented yellow and pink flowers. It likes a cool root run. *Rosa* 'Little Rambler' (syn. *R.* 'Chewramb') is a miniature climber with sweetly scented blush-pink flowers in summer. The stems are pliant and can be trained around a window in a fan shape.

Lilies offer a rich and exotic perfume, and *Lilium* Pink Perfection Group are trumpet-flowered lilies with deep pink flowers. *Philadelphus* 'Manteau d'Hermine' (mock orange) bears sweetly scented double creamy-white blooms. The aromatic foliage of *Myrtus communis* (common myrtle) and certain pelargoniums widen the types of perfume considerably. The myrtle is a neat evergreen shrub with glossy leaves, aromatic when crushed, and clusters of white flowers. The scented-leaved pelargoniums are compact, frost-tender perennials with leaves that are strongly scented when broken between the fingers.

Annuals such as *Lathyrus odoratus* (sweet pea), a hardy climber available in a wide range of colours, and stocks are easy and inexpensive to grow, and they produce some delicious scents into the bargain. *Malcolmia maritima* (Virginian stock), a hardy annual, has clusters of mauve or pink flowers. *Matthiola incana* 'Cinderella' is a dwarf variety of annual stock that has columns of colourful flowers.

1 *Lonicera periclymenum* 'Belgica'
2 *Rosa* 'Little Rambler'
3 *Myrtus communis*
4 *Lilium* Pink Perfection Group
5 *Lathyrus odoratus*
6 *Philadelphus* 'Manteau d'Hermine'
7 *Pelargonium* Fragrans Group
8 *Matthiola incana* 'Cinderella'
9 *Malcolmia maritima*

Opposite: In the shadow of the scented trumpets of *Lilium* Pink Perfection Group, the prostrate foliage of creeping thyme (*Thymus pseudolanuginosus*) moulds itself attractively around the shaped rim of a large pot.

Nothing can beat annuals for a colourful summer show. Here pink stocks bring rich fragrance to a planting that also includes red geraniums and yellow mimulus.

This way you can avoid the risk of starting right from scratch with seed, but it is considerably cheaper than buying fully grown plants in late spring and early summer.

Tender perennials, as their name suggests, go on for years but will not tolerate frost. Just a handful are scented, but what wonderful plants they are. Chocoholics will adore *Cosmos atrosanguineus* (chocolate cosmos) – the blooms are like little dark red dahlias that have the mouth-watering

Easily raised from seed, annual sweet peas should be grown in deep pots and can be trained up a variety of supports, in this case nothing more than a few tall sticks.

Top 10 Scented Annuals

HA = hardy annual
HHA = half-hardy annual

- *Asperula orientalis* (blue woodruff) HA
- *Centaurea moschata* (sweet sultan) HA
- *Ipomoea alba* (moonflower) HHA
- *Lathyrus odoratus* (sweet pea) HA
- *Lobularia maritima* (sweet alyssum) HA
- *Matthiola incana* (stock) HHA
- *Matthiola longipetala* (syn. *M. bicornis*; night-scented stock) HA
- *Nicotiana alata* (tobacco plant) HHA
- *Reseda odorata* (mignonette) HA
- *Zaluzianskya capensis* (night phlox) HA

smell of dark chocolate – while *Heliotropium* (heliotrope, cherry pie) is almost as good. *Mirabilis jalapa* (marvel of Peru) grows from tubers – these are often available in spring – and flowers in the afternoon and evening, hence its other common name, the four o'clock plant. These three are all upright and bushy, while *Nemesia denticulata* (syn. *N. d.* 'Confetti') and *N.* 'Fragrant Cloud' have a lax, spreading habit and are lovely for wall baskets next to a window, where the sweet scent of the pale pink blooms can drift indoors. For exceptionally fragrant foliage there are pelargoniums with scented leaves (see page 51).

Above: Pale pink *Nemesia denticulata* (syn. *N.d.* 'Confetti') has a light, spicy fragrance and blooms over a long period.

Right: Ideal for those out at work during the day, *Mirabilis jalapa* (four o'clock plant) has scented blooms that only open in the late afternoon and evening.

A Sweet Pea Tower

THESE splendid, speedy and easy-to-grow climbers scramble up by means of tendrils to reach a height of around 1.8m (6ft). A magnificent display can be made by growing a dozen or so plants in a large pot, trained up a support to make a sweet pea tower. Throughout summer, long stems bearing a number of small, fluted, very fragrant flowers are produced in succession until autumn.

Sow sweet peas in autumn or spring. Use deep pots to accommodate their long roots or use root-trainers, which have grooved sides to encourage a good root system. Soak the seed in tepid water for a few hours before sowing to soften the hard seed coat. Grow the plants on in a greenhouse or cold frame – heat isn't necessary.

For a sweet pea tower choose a large, deep container because sweet peas are deep-rooting and hungry plants, needing plenty of good compost. Put crocks or stones in the base of the pot to aid drainage and fill it two-thirds full with potting compost. Set up a wigwam of 2.1m (7ft) bamboo canes, putting the bases firmly in the pot and tying the tops together. For a neat finish buy a special cane holder to secure the tops. Plant the sweet peas, topping up the compost as you go.

For best results sweet peas need regular tender loving care. Tie in new stems every few days, using string or twine. Never allow the compost to dry out. Feed weekly with a liquid fertilizer that is high in potash, but take care not to overfeed, or leaves will be produced at the expense of flowers. Deadhead every few days as soon as blooms fade.

Which variety?

Although most sweet peas are fragrant, some are much more highly scented than others. Old-fashioned varieties have smaller flowers than many newer types but with a much stronger perfume, and there are now a number of newer varieties that have been bred with scent in mind too. A huge range of colours is available in just about every shade except yellow. Here are just a few of the many good varieties.

Modern varieties

'Cambridge Blue' – light blue
'Daphne' – lilac-mauve
'Queen Mother' – salmon pink
'Southbourne' – pale pink
'Windsor' – dark maroon

Old-fashioned varieties

'Janet Scott' – shell pink
'Flora Norton' – clear blue
'Lord Nelson' – navy blue
'Mrs Collier' – cream
'Painted Lady' – red and white
'Matucana' – blue and purple

Pots of Cherry Pie

HELIOTROPIUM (heliotrope) is an old Victorian favourite, often known as cherry pie, which looks good right through summer, and several different varieties can be grouped together for a feast of fragrance. Large domed or flattened flowerheads are composed of many tiny blooms, which have a deliciously sweet scent, and they are wonderful for attracting butterflies. The foliage is handsome, too, with oval to lance-shaped dark green, occasionally purple-tinged leaves, which are deeply ridged.

Old-fashioned varieties invariably have the best scent. *H.* 'Chatsworth' is deep mauve; *H.* 'President Garfield' is mauve and blue; *H.* 'White Queen' is pale mauve fading to white; and *H.* 'White Lady' is a compact form with white flowers that are pink in bud.

Heliotrope can also be trained as a standard and used as the centrepiece of a group of plants. Ready-grown standards are sometimes available, although it is much cheaper and more rewarding to grow your own over a period of two or three years. Start with a healthy cutting and tie it to a cane. As it grows, remove the leaves on the lower part of the stem until it has reached the desired height – 60–90cm (2–3ft) is ideal. Pinch out the tip, which will encourage the production of side shoots. In turn, pinch these out when well grown to encourage a bushy head to form.

For best results, grow heliotrope in a sheltered spot in full sun. Overwinter plants under cover in a frost-free environment. Alternatively, take cuttings in midsummer to overwinter on a windowsill.

1 *Heliotropium* 'Chatsworth'
2 *H.* 'President Garfield'
3 *H.* 'White Queen'
4 *H.* 'White Lady'

Scented-leaved Pelargoniums

A great favourite with the Victorians, these fascinating plants are coming back into fashion as today's gardeners discover the delightful leaf fragrances, such as rose, peppermint, lemon, balsam and orange. The flowers are nowhere near as showy as those of bedding geraniums, but when plants smell this good, who cares? In any case, the leaves offer subtle yet excellent ornamental value, with many different shapes, textures and shades of green, as well as some variegated forms. To achieve a supremely handsome plant, grow singly in a pot that is about 25–30cm (10–12in) across and pinch out the shoot tips to encourage a bushy habit.

Pelargoniums are easy to grow, provided they have sun, well-drained compost and not too much water. Overwintered plants should be kept on the dry side in a frost-free greenhouse or conservatory. Cuttings taken in late summer root easily and can be grown indoors on the windowsill as insurance against losing the parent plant. Some of the best of the many varieties and their scents are listed here.

P. 'Attar of Roses' – rose
P. 'Copthorne' – cedar
P. crispum 'Variegatum' – lemon
P. denticulatum 'Filicifolium' – balsam
P. Fragrans Group – pine
P. 'Graveolens' – lemon
P. 'Lady Plymouth' – lemon
P. 'Lilian Pottinger' – pine
P. 'Mabel Grey' – lemon
P. odoratissimum – green apples
P. 'Prince of Orange' – orange
P. tomentosum – peppermint

The foliage of pelargoniums is attractive in its own right or used as a contrast with other plants. Variegated Pelargonium 'Lady Plymouth' looks stunning with *Fuchsia* 'Thalia'.

Up, Up and Away – Climbing Plants

Gardening on the vertical offers some great opportunities to cram in more plants where space is limited, but there are other benefits. A pillar of flowers (see page 49) or a climber trained on a trellis fan (see page 71) makes an excellent centrepiece to a group of containers and introduces some height into areas where it is often sorely needed, and vigorous climbers can be trained over an arch or arbour. On a practical level, climbing plants can be strategically placed to provide some privacy or to screen an ugly object from view. In many places, such as around the house, the ground may be entirely paved or concreted and plants in containers may be the only way to clothe walls and fences with a beautiful cloak of flowers and foliage.

Perennial climbers are excellent because little work is involved apart from a spring top-dressing, and you will have a substantial plant that will be there for years. Do use a container that is sufficiently large, such as a wooden half-barrel, or within a couple of years the plant will become potbound and start to decline. *Jasminum officinale* (common jasmine) is easy to grow and vigorous, needing plenty of room to climb, and it is soon covered in clusters of pure white, coconut-scented flowers against dark green leaves. For a more decorative appearance, try *J. o.* 'Fiona Sunrise', with leaves that are brightly suffused with gold. *Trachelospermum* spp. bears jasmine-like blooms, although this classy evergreen needs a little more care. Grow it in a sunny, sheltered site – up a wall is fine as its stems are self-clinging – and this first-rate plant will reward you with lots of deliciously scented flowers. *Lonicera* spp. (honeysuckle) is as untidy as trachelospermum is neat, but the much-loved fragrance of its long-lasting flowers makes it an almost essential ingredient of the scented garden. Although it is not ideally suited to growing in a container, honeysuckle will do well if it is sited in a partially shaded site, preferably with its roots shaded further by other containers so that they remain cool and moist at all times. Don't be afraid to give honeysuckle a really hard haircut in early spring in order to achieve a tidier plant that flowers over a longer period.

Below: Vigorous climbers can be grown in large containers for a profusion of flowers and foliage. Plant *Jasminum officinale* (summer jasmine) to grow over a wall, arch or pergola to enjoy its potent perfume.

Creating a Fragrant Evening Garden

1 *Jasminum officinale*
2 *Choisya ternata* 'Sundance'
3 *Lilium* 'Star Gazer'
4 *Nicotiana*
5 *Mirabilis jalapa*
6 *Matthiola longipetala*
7 *Zaluzianskya capensis*

SCENTED plants add a wonderful extra dimension to the evening garden in summer and are particularly enjoyed by many people who see their garden only at this time during the working day. As the day reaches its end, there's nothing better than sitting down to relax and breathe in some glorious perfumes while listening to the garden coming alive at night.

Several plants only open their flowers or release their scent in the evening, while the perfumes of other plants intensify considerably at this time. Choose a sheltered corner for your evening garden, west-facing if possible so that you can enjoy the last of the sun. Evening-scented plants are also excellent positioned around your house doors and windows, where their fragrance can drift inside on warm evenings.

Jasminum officinale (summer jasmine) can be grown in a large container and allowed to twine up a wall or over an arch. Clusters of white, coconut-scented flowers show up beautifully against dark foliage. *Choisya ternata* 'Sundance' (Mexican orange blossom) has star-shaped white, sweetly scented flowers that smell much stronger in the evening. The attractive glossy, golden leaves give off a strong scent when they are bruised.

Lilies smell much stronger in the evening; see page 75 for a list of scented varieties.

Nicotiana (tobacco plant) has a delicious, strong perfume, and although there are many colours available, white flowers have by far the best scent.

Mirabilis jalapa (marvel of Peru, four o'clock plant) is the perfect plant for commuters because the multi-coloured flowers don't open until late afternoon.

Matthiola longipetala (syn. *M. bicornis*; night-scented stock) is not the neatest of plants, but the tiny lilac-coloured flowers have a potent clove scent. *Zaluzianskya capensis* (night phlox) has starry white flowers that open in the evening to release a strong fragrance.

In addition to the above plants that flower in summer, *Hesperis matronalis* (dame's violet, sweet rocket) is an easily grown biennial with mauve or white flowers, which blooms in late spring. Untidy in habit and liable to flop, but irresistible for all that, it's best tucked towards the back of a group of plants.

Cut Flowers

- Little black beetles infest many flowers, particularly sweet peas, and can be annoying when they fly all over the room from blooms brought inside. Remove them by placing the bunch of flowers (in water) in a cool shed or garage that has a window. In the course of a couple of hours, the beetles will leave the flowers and head for the light.
- For long-lasting cut flowers in summer, pick blooms early, before the day really warms up. Use sharp scissors or secateurs to make a slanting cut across the base of the stem to expose the maximum area for taking up water.
- Before taking the flowers indoors into vases, give them a good drink first by standing the blooms for a couple of hours up to their necks in a bucket of water.
- Add some cut-flower food to the vase water to prolong the life of the blooms.

Opposite: Shrubby herbs like French lavender (far right) last for years with little maintenance. This row of small pots makes the most of a brick-edged raised bed.

When you are buying, take care to avoid the few species that have no scent, in particular *Lonicera* × *brownii* (scarlet-trumpet honeysuckle), *L.* × *tellmanniana* and *L. tragophylla*.

Annual climbers make first-rate seasonal gap-fillers, either in pots on their own or popped in with one of these perennial climbers for extra colour. *Lathyrus odoratus* (sweet pea) is an absolute must for every garden, particularly as sweet peas are quick, cheap and easy to grow (see page 49). *Ipomoea alba* (moonflower), a half-hardy annual, needs a bit more care and also prefers a sunny, sheltered site, but it is well worth the effort for its superb, saucer-shaped, pure white flowers, which give off a rich perfume in the evening. Annual climbers make good partners for miniature climbing roses too (see page 57).

Perennial Perfumes – Shrubs and Herbaceous Perennials

There are several shrubby 'must-haves' for the scented summer garden, starting with *Philadelphus* spp. (mock orange), which bears masses of white, wonderfully perfumed blooms. Many gardeners are familiar with the tall-growing varieties such as *P.* 'Virginal' – this old faithful is out of bounds for a pot – but there are a couple of compact varieties that are fine in containers. *P.* 'Manteau d'Hermine' has double blooms, while those of *P. microphyllus* are single. Flowers are the main attraction because philadelphus is, frankly, dull for the rest of the year.

The same cannot be said for *Choisya ternata* (Mexican orange blossom), which has beautiful, glossy, evergreen, lobed leaves. These are deeply cut in *C.* 'Aztec Pearl', while those of *C. ternata* 'Sundance' are bright yellow. Stick to one of these two rather than the species, *C. ternata*, which is rather too vigorous for a pot. *C. t.* 'Sundance' makes a lovely backdrop for deep blue lavender such as *Lavandula angustifolia* 'Hidcote', a classic cottage-garden plant, or the slightly tender French lavender (*L. stoechas*), which has unusual flowers topped with a pair of winged bracts. Add some coarse grit to the potting compost because lavender needs sharp drainage. *Myrtus communis* subsp. *tarentina* (myrtle) has glossy, aromatic foliage and fragrant flowers that are white-petalled with a pin-cushion-like tuft of stamens. This variety is hardier than the species *M. communis*, although it is best given winter protection in cold areas to be on the safe side. Variegated varieties are even more decorative but definitely need to go under cover for the winter.

Perennials look delightful when they are grown individually in small pots that can then be grouped around these larger plants. *Dianthus* (pinks) have a wonderfully rich clove scent, and there is a huge range to choose from. Modern hybrid pinks like *D.* 'Doris' bloom pretty much right through the summer, while old-fashioned varieties generally bloom just once, in midsummer, but include some forms with beautifully

Left: Cottage-garden favourites, *Dianthus* (pinks) have clove-scented flowers and come in an enormous range of varieties.

marked and laced flowers. Some date as far back as Tudor times, like the dark red and white *D.* 'Sops-in-wine' which was actually used then to flavour wine.

Pinks go well with the lavender-blue flowers and aromatic grey-green leaves of *Nepeta* × *faassenii* (catmint), which, as the name suggests, is irresistible to cats – not a plant to grow if you want to keep the neighbourhood's feline population away. Lastly, and making a good contrast of shape to both the above perennials, is the sumptuous *Iris graminea* (plum-tart iris), which has small, reddish-purple flowers that have the most delicious and mouth-watering scent.

Fragrant Favourites – Roses

With long-lasting, attractively shaped and often fragrant blooms, roses are immensely popular summer flowers. Although the great majority are not suitable for containers simply because their root systems grow too deep, there are two types of rose that are suitable for growing in large pots. These are patio climbers and patio bush/miniature roses.

Miniature or 'patio' climbers form a comparatively new group of roses that reach only about 1.8m (6ft) and can be grown in a large tub such as a half-barrel for training up a wall, fence or a piece of trellis.

Opposite: A shrub makes a good centrepiece for a large container. A tall variegated myrtle (*Myrtus communis* 'Variegata') has attractive and aromatic foliage, plus white, fragrant flowers.

To encourage maximum flower production, train the stems in a wide fan-shape as this restricts the flow of sap along the stems and thus boosts flowering. *R.* 'Little Rambler' (syn. *R.* 'Chewramb') is an excellent scented variety, producing clusters of small, very pale pink blooms right through the summer. It has the pliant stems that are typical of a rambler rose and as such is easy to train. *R.* 'Nice Day' (syn. *R.* 'Chewsea') bears double blooms that are soft salmon pink and have a sweet fragrance, while *R.* 'Warm Welcome' (syn. *R.* 'Chewizz') has neat, well-formed orange blooms that are set off well against coppery-green foliage.

Unlike the majority of roses, dwarf bush or patio roses thrive in large pots. Comparatively few are scented, however, though the pretty and long-lasting blooms of *Rosa* 'Sweet Dream' have a light fragrance.

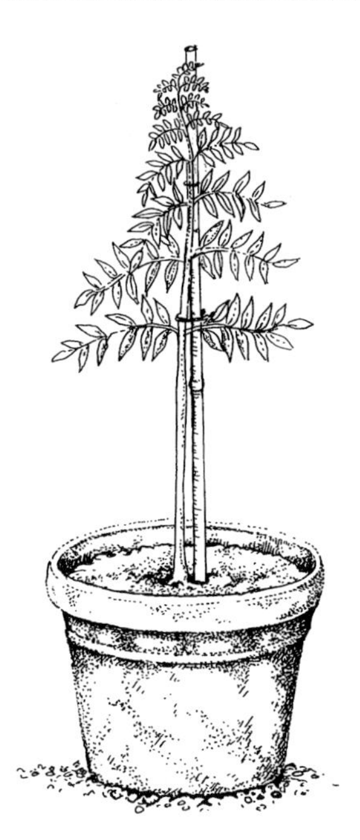

1 Use a large pot made of a heavy material such as terracotta. Stake with a stout bamboo cane.

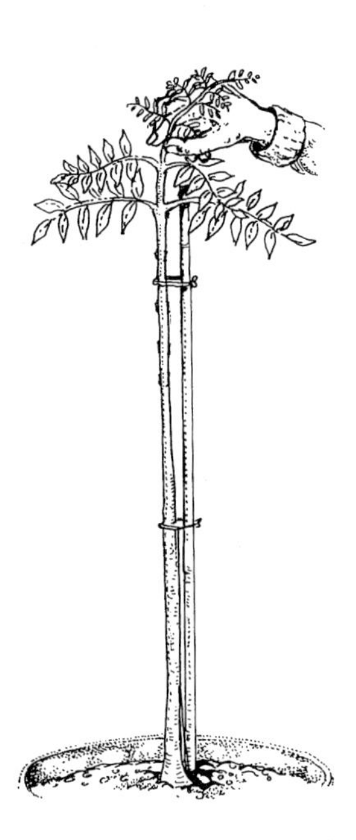

2 Pinch out the top shoot when it is about 1.5m (5ft) high. Allow several branches to develop.

Training a Standard Wisteria

A VIGOROUS climbing plant, wisteria is frequently seen draped over the entire side of a house and festooned with huge racemes of deliciously scented flowers in late spring. Few people realize, however, that this delectable plant can be grown in a large container and trained as a 'standard' – that is, with a short, bare stem and a head of flowers and foliage.

Wisteria floribunda (Japanese wisteria), the most compact type, offers a range of colours, including blue, purple, peach, pink and white. Buy plants from a reputable source and check that they have been produced by grafting; plants raised by other methods can take a number of years to flower. Choose a plant that has an obvious leading shoot. You will need a large, heavy container, and do not use a plastic pot because the plant will ultimately become top heavy. Put a layer of crocks in the base of the pot and plant the wisteria using a soil-based potting compost. Tie the leading shoot to a stout cane – this stem will eventually form the 'trunk' of the plant – and remove the lower shoots. Continue doing this as the plant grows upwards until it has reached a height of 1.5–1.8m (5–6ft), and then pinch out the growing point. Let shoots develop to form the 'head' and pinch out any shoots remaining on the supporting stem. As the head reaches the size you want, trim it to shape using secateurs.

Wisteria needs pruning twice a year. In midsummer, shorten side shoots to five or six buds and pinch out the growing tips of main shoots if they are becoming over-long. In midwinter, cut the pruned side shoots again, this time to two or three buds.

Your standard wisteria can be underplanted with a low-growing, shade-tolerant plant if you wish. Annuals are best, because permanent plants could compete with the wisteria. *Asperula orientalis* (blue woodruff) is a fragrant annual for summer flowers and, unlike most annuals, thrives in shade. Trailing lobelia is an unscented alternative.

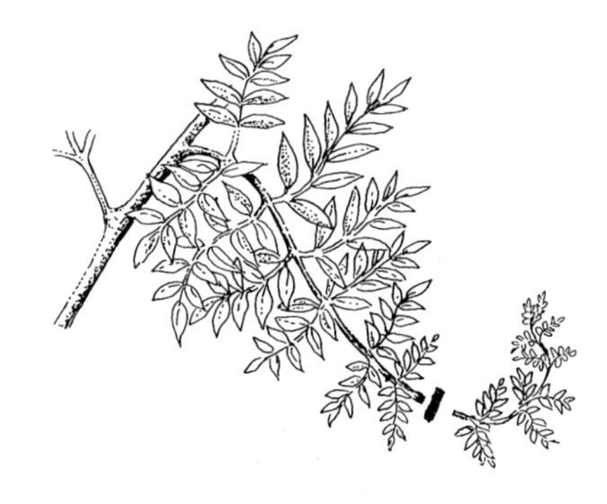

3 In summer, shorten side shoots to five or six buds from the main stem.

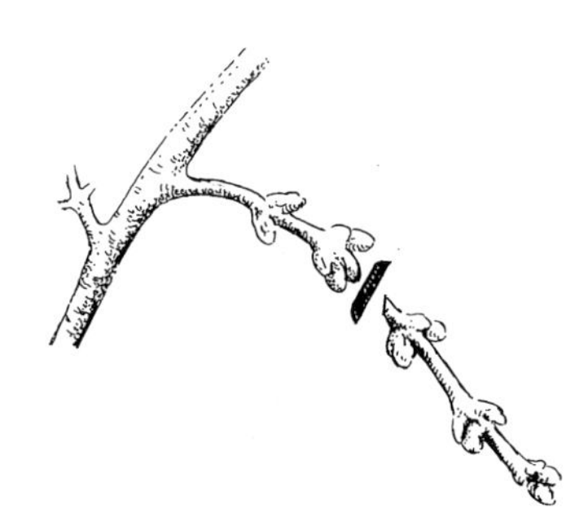

4 In winter, cut the side shoots back further to two or three buds.

A fragrant and much-loved plant, the wallflower (*Erysimum cheiri*) adapts to any surroundings, from casual cottage gardens to this grand stone container.

Miniature, patio and ground-cover roses reach a maximum of 60–90cm (24–36in) high and are the only bush type that can sensibly be grown in a pot. These are mostly of modern introduction and, alas, many are very good-looking but with no scent to speak of. There are a few exceptions, including *R.* 'Norfolk' (syn. *R.* 'Poulfolk'), which has bright yellow, double, very fragrant flowers, *R.* 'Sweet Magic' (syn. *R.* 'Dicmagic'), which bears profuse quantities of fragrant, orange and gold blooms on a compact bush, and *R.* 'Yorkshire', which produces small clusters of white, semi-double flowers. Fortunately, the tide of opinion is now turning in favour of perfumed roses, and so we can look forward to some fragrant new introductions. (See page 120 for varieties.)

Seasonal Reminders

Propagation

- In early summer sow seed of annual flowers to bloom later in the season, which will follow on from plants raised from spring sowings.
- Biennials, such as wallflowers, sweet williams, sweet rocket and certain varieties of stocks, can be sown in a nursery bed by early summer.
- Most shrubs and climbers can be propagated by half-ripe cuttings or by layering.
- *Dianthus* (pinks) can be propagated by 'pipings' – cuttings taken by pulling off shoot tips with several pairs of leaves.

Cutting Back on Maintenance

Containers are lovely but labour-intensive, because the plants rely on you for all their needs. Regular watering and feeding are essential, but there are some good ways to lighten the load.

An irrigation system is an excellent investment that will save you many hours of watering. A micro-drip system can be set up in a few hours and it will be there for years. Add a water timer, and your irrigation becomes fully automatic, so your plants are even taken care of during holidays and busy working days. Such a system is better for your plants too, because it can be programmed to water in the early morning and late evening so there is no danger of wet foliage being scorched by bright sunlight. The slow, gentle drip of water from irrigation nozzles is much better than a great gush from a hosepipe. (See panel on page 101 for further details.)

All container plants need regular feeding during the growing season. Potting compost has enough nutrients to last for about six weeks, but after that you'll need to top it up. Rather than time-consuming weekly liquid feeds, add some controlled-release fertilizer to the compost. This is 'smart' fertilizer, which is coated with temperature-sensitive resin so that nutrients are released only when it is warm enough for the plant to grow and take them up. It will last for the whole summer, although seasonal flowering plants do benefit from a couple of top-up liquid feeds towards autumn.

- In late summer, take cuttings of tender perennials, such as nemesia and scented-leaved pelargoniums, to overwinter on windowsills.

Pruning

- Trim herbs two or three times during the growing season, to ensure compact plants with lots of attractive young growth.
- Lavender that has flowered should be trimmed lightly using shears.
- Wisteria should be pruned in midsummer.
- Spring- and early-summer-flowering shrubs and climbers, such as *Akebia*, *Philadelphus* and *Syringa* (lilac), should be pruned after flowering.

General

- Water containers frequently.
- Feed weekly with a liquid fertilizer or apply a one-off dose of controlled-release fertilizer.
- Deadhead flowering plants regularly to prolong flowering.
- Inspect plants frequently for signs of pests or diseases. Try to resolve the problem without resorting to chemicals, either by picking and squashing infestations of insects in the early stages or through the use of a biological control.
- Towards the end of the season, stock up on bulb catalogues and place your orders promptly. Many bulbs – particularly those that flower in winter or early spring – should be planted by early autumn

Autumn – Aromatic Splendour

As the nights begin to draw in, the garden seems to take on an awareness that its days of plenty are numbered and gathers its resources for a final, glorious feast of colour and fragrance. Cooler mornings and an increase in the dew-fall intensify the garden's scents when the sun's warmth arrives. Aromatic foliage, which has often been eclipsed by the masses of scented flowers during summer, comes into its own now with herbs and other plants providing an increasing fall-back as seasonal plants come to an end. This is a time of harvest too, for cutting, drying and freezing herbs for winter use. But autumn is also a time of planning and preparation for next year, planting bulbs and biennials and sowing seed to ensure there will be plenty to look forward to in the coming year.

Autumn is a time to plan ahead for the following year's fragrance. The red-berried *Skimmia japonica* subsp. *reevesiana* (centre) bears clusters of scented white flowers in spring.

A Group of Pots for Autumn

1 *Clematis rehderiana*
2 *Clematis flammula*
3 *Lathyrus odoratus*
4 *Foeniculum vulgare* 'Purpureum'
5 *Reseda odorata*
6 *Asperula orientalis*
7 *Mentha* × *piperita* f. *citrata*
8 *Salvia officinalis* Purpurascens Group
9 *Nicotiana*
10 *Petroselinum crispum*
11 *Matthiola longipetala*
12 *Calamintha grandiflora*

THE garden's bounty may not be so prolific now, but fragrant flowers still abound if the right varieties are chosen. Many annuals, such as mignonette and tobacco plant, keep blooming well into autumn, as will late-sown sweet peas, stocks and blue woodruff.

Two lovely late-blooming clematis offer a profusion of scent and colour. *Clematis rehderiana* is a vigorous climber bearing dangling, tubular, pale yellow flowers that have a scent of cowslips. *C. flammula* is also vigorous, and it bears clusters of starry white, strongly scented flowers.

The hardy annual *Lathyrus odoratus* (sweet pea) is an easily grown climber, available in a wide range of colours. Autumn-flowering plants can be obtained from a late spring sowing.

Herbs really come into their own now, with aromatic leaves that give off a wide variety of scents when crushed. *Foeniculum vulgare* 'Purpureum' (bronze fennel) is a tall perennial herb with attractive, feathery, purple-tinged foliage that is aromatic when crushed. *Mentha* × *piperita* f. *citrata* (eau-de-cologne mint) is a spreading perennial, with rounded, dark green leaves that smell strongly when crushed, and *Salvia officinalis* Purpurascens Group (purple sage) is a rather shrubby perennial with purple, fragrant foliage. *Petroselinum crispum* (parsley) has curled, dark green leaves, while *Calamintha grandiflora* (large-flowered calamint) is a low-growing herb with aromatic leaves.

Among the hardy annuals are *Reseda odorata* (mignonette), which has heads of yellow-green to white, fragrant flowers, and *Asperula orientalis* (blue woodruff), which has finely cut leaves and heads of pale blue, scented flowers. Because it is tolerant of shade, blue woodruff is useful for underplanting. *Matthiola longipetala* (syn. *M. bicornis*; night-scented stock) has small, very fragrant flowers; sow seed in late spring to early summer for autumn flowers.

Nicotiana (tobacco plant) is a half-hardy annual, and the scent of its large flowers is particularly strong at night.

Beauty and Practicality – Herbs

Above: Aromatic foliage can be enjoyed all year but comes into its own in autumn when most flowers are fading. As a plant to clip into all manner of shapes, *Laurus nobilis* (sweet bay) reigns supreme.

Right: Herbs look good planted together in a large pot so that their colours and shapes contrast. Bay, marjoram and curry plant are all perennial but need regular pruning to maintain a balance of growth.

Opposite: Large wooden barrels provide a lot of root space. This one is packed with herbs including pink lavender and variegated lemon balm.

Many herbs look very attractive in pots and are easy to grow – in addition, there is something immensely satisfying about harvesting and eating something that is home grown. Most herbs have leaves that are delightfully aromatic when crushed or brushed against, and the flowers of some are attractive to butterflies and bees. Some varieties of certain herbs are much more ornamental than others, and these decorative ones are obviously most attractive for containers that occupy a high-profile site. For example, the golden-variegated form of lemon balm is much more decorative than the plain green form; similarly, purple-leaved fennel is more attractive than green fennel. These ornamental forms are, in the main, those that are included in the Plant Directory on page 117.

Container-grown herbs have several advantages over those that grow in the ground. Top of my list is convenience, for when I'm in the middle of cooking it's very handy just to reach outside the kitchen door

A Herb Hanging Basket

HERBS are immensely decorative as well as useful, and for the ultimate in convenience plant a hanging basket of herbs to grow as near to the kitchen door as possible. The selection of herbs described here does well in a partly shaded spot.

For a 30cm (12in) open-mesh hanging basket you will need: material for lining; six *Petroselinum crispum* (parsley); three *Origanum vulgare* 'Aureum' (golden marjoram); three *Mentha pulegium* (pennyroyal); and three *Allium tuberosum* (garlic chives). The parsley plants should be fairly small, because they dislike root disturbance. For extra colour you could sow a few seeds of *Tropaeolum* (nasturtium), all parts of which are edible and can be used to add a warm, peppery flavour to salads.

Stand the basket in a large pot or bucket for stability and line it, using wool, card, coir or moss. Fill it about one-third full of compost, and put in the pennyroyal and three of the parsley plants, spacing them evenly around the sides of the basket and squeezing the rootballs slightly in order to get them through the mesh. Put in another third of the compost and plant the remaining herbs, with the parsley in the centre and the marjoram and garlic chives around the edge.

To aid future watering, when the plants will be well established, sink a small plastic pot in the top of the basket so that only the rim is visible. If you simply fill the pot with water, it will go straight to the roots instead of running off the top.

Fill the spaces between the plants with compost and water well, using a watering can fitted with a fine rose. If any gaps subsequently appear, top up with compost.

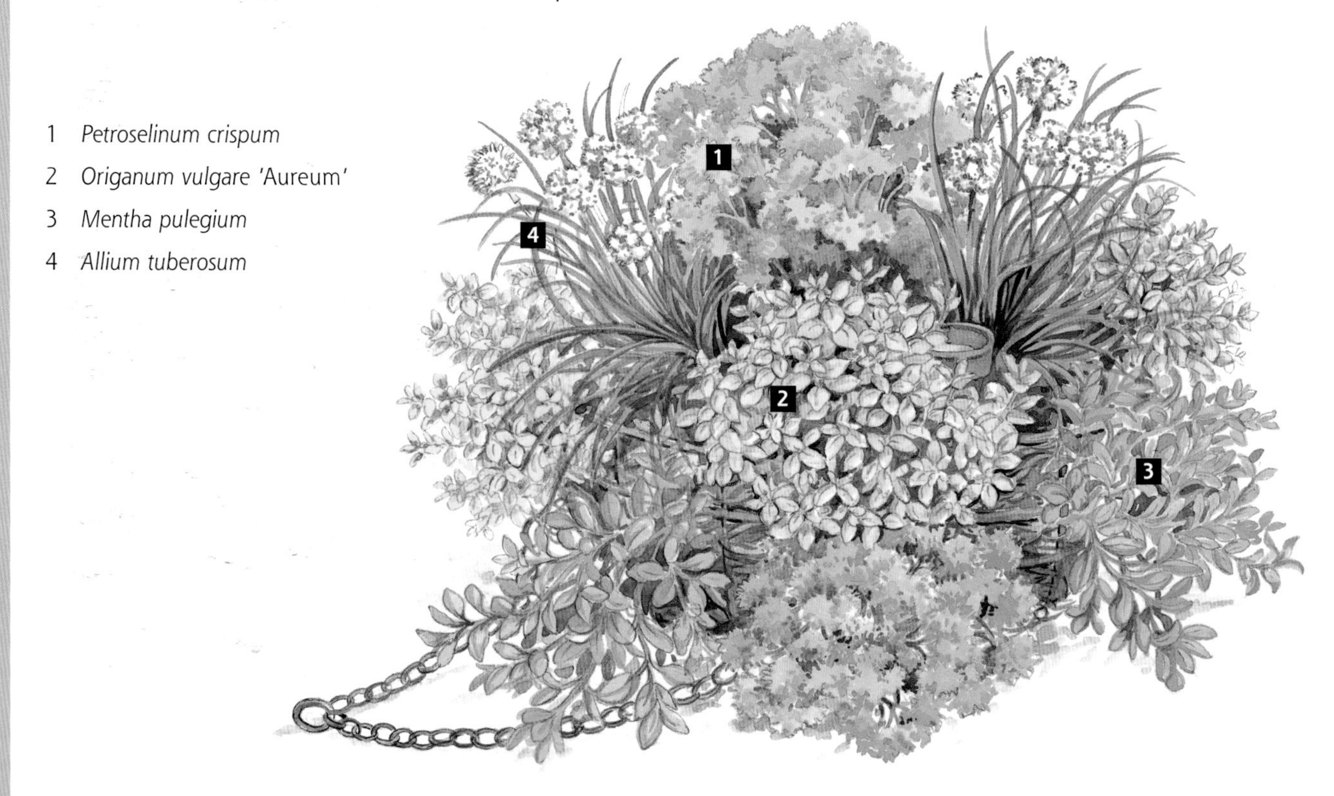

1 *Petroselinum crispum*
2 *Origanum vulgare* 'Aureum'
3 *Mentha pulegium*
4 *Allium tuberosum*

for a few sprigs of foliage, as opposed to running down the garden in the dark! Fresh-picked herbs have maximum flavour and goodness too, and the foliage also stays cleaner, particularly if the pots are topped with a layer of fine gravel.

For practical growing purposes, herbs can be divided into several groups. Shrubby types, such as *Laurus nobilis* (bay), *Helichrysum italicum* subsp. *serotinum* (syn. *H. serotinum*; curry plant) and *Rosmarinus officinalis* (rosemary), are evergreen and increase in size from year to year.

Mouth-watering Mints

There is far more to mint than just the garden variety of mint-sauce fame. Try a few of the varieties listed below for a wonderful range of scents. Most mints are very invasive, spreading rapidly by underground runners, so I do advise growing these plants in individual pots.

- *Mentha* × *gracilis* 'Variegata' (ginger mint) – leaves are variegated green and gold with jagged edges; a delicate, warm fragrance.
- *M.* × *piperita* (peppermint) – pointed, dark green, purple-tinged leaves with a strong peppermint scent. Black peppermint has much darker leaves.
- *M.* × *piperita* f. *citrata* (eau-de-cologne mint) – large, rounded, dark green leaves with a powerful eau-de-cologne fragrance. Use to perfume bathwater rather than in cooking.
- *M.* × *piperita* f. *citrata* 'Basil' (basil mint) – green, red-tinged leaves with a sweet yet spicy scent. Good with pasta.
- *M. pulegium* (pennyroyal) – a ground-covering variety with tiny leaves that have a peppermint scent. Do not consume if pregnant. Grow near the door to keep ants away.
- *M. spicata* (spearmint) – garden or common mint, with pointed green leaves, good for mint sauce.
- *M. spicata* 'Moroccan' (Moroccan mint) – a bushy plant with pale green leaves and a light mint scent. Excellent for yogurt-based sauces.
- *M. suaveolens* (syn. *M. rotundifolia*; apple mint, woolly mint) – rounded, cream-and-green leaves with a light, fresh scent.

Mint is invasive and is best grown on its own. Group with other individual plants – sweet bay and a scented pelargonium are shown here – to create an attractive display.

Herbaceous perennials, such as *Allium schoenoprasum* (chives), *Foeniculum vulgare* 'Purpureum' (bronze fennel) and *Mentha* spp. (mint), die back to the ground in autumn and regrow in spring. Annuals and biennials such as *Petroselinum crispum* (parsley) and *Anethum graveolens* (dill) need to be grown from seed every year.

Ocimum basilicum (basil) is a half-hardy annual that needs slightly different treatment. It should be raised under cover in a warm environment and moved outside to a sunny, sheltered spot once all danger of frost is past. This outstanding plant offers a real taste of the Mediterranean, and its strongly aromatic leaves are superb for flavouring many dishes. A number of different varieties of sweet basil are available, including the purple-leaved forms *O. b.* 'Dark Opal', *O. b.* 'Purple Ruffles' and *O. b.* 'Red Rubin', which can be used to make a dramatic contrast with other plants.

The easiest way to grow herbs is in individual containers, the size of which should be varied to match the size and vigour of each plant. Alternatively – or as well – grow several herbs together in a large pot to make an attractive display, although it will be necessary to trim the plants regularly to ensure a vigorous variety doesn't swamp its companions. Such a combination must also be carefully chosen to make sure that they like the same conditions – for instance, that all sun-lovers are planted together.

To create a potted herb garden, choose a variety of plants with contrasting shapes and colours as well as a range of sizes. A bay tree would make an excellent centrepiece. Either grow your own specimen from a small plant, or you could splash out on a splendid ready-grown 'lollipop'

Freezing Herbs

Drying is the traditional way of storing herbs for winter, but freezing is much easier. Simply take an ice-cube tray and snip the washed leaves into it, pressing them down very lightly so each section is about two-thirds full. Top up the tray with water and put in the freezer. After a few hours, when the cubes are fully frozen, empty them out into a polythene freezer bag and label the contents. Simply add a cube or two to your soup, sauce or casserole as required.

Opposite: Herbs, with their cottage-garden image, are perfectly at home when placed higgledy-piggledy around the garden.

Left: Neat-growing herbs such as thymes can be grown in small pots and used in a restrained grouping, as with these *Sempervivum*.

or pyramid. Next in size come tall herbs, particularly fennel, which quickly develops upright stems clothed with beautiful feathery foliage, and rosemary (*Rosmarinus officinalis* 'Miss Jessopp's Upright'). Then there are medium-sized shrubby herbs such as *Helichrysum italicum* subsp. *serotinum* (curry plant), *Salvia officinalis* (sage) and *Satureja montana* (winter savory). All their different shapes contrast beautifully with prostrate or spreading growers such as *Thymus* spp. (thyme), *Melissa officinalis* (lemon balm), *Origanum* spp. (marjoram) and *Mentha* spp. (mint).

One herb I really would not recommend is *Ruta graveolens* (common rue). The lobed, blue-green foliage may look very handsome, but it gives off a perfectly revolting smell, and contact with the leaves may also cause a very severe allergic skin reaction. The foliage has most potential to cause damage in strong sunlight or immediately after rain.

Herb Care

- Herbs need good drainage. With the exception of suspended containers, use soil-based compost with added grit and raise pots just off the ground to let water flow away easily.
- Avoid overfeeding, which encourages too much lush, leggy growth. Potting compost contains enough nutrients for a couple of months; after that, apply a liquid feed once every two or three weeks.
- Shrubby herbs are susceptible to frost damage in cold areas. Ideally, move pots into an unheated porch or greenhouse for the winter. Next best is to move them to a sheltered spot and protect them with horticultural fleece and bubble polythene (see pages 100–101).
- Cut back the dead foliage of perennial herbs in spring. It may look untidy through the winter but it does give extra frost protection.
- Potbound perennial herbs can be rejuvenated by dividing and replanting in early spring.
- Trim herbs frequently to encourage plenty of fresh growth and a compact plant.

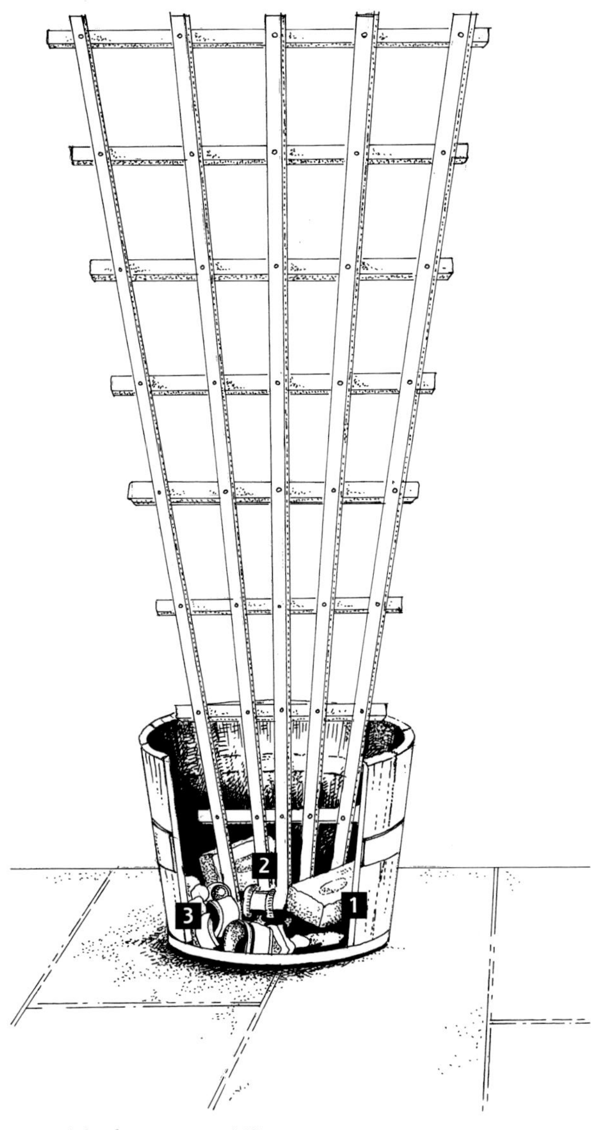

1 Bricks for extra stability

2 Wire threaded through drainage holes and tied to bottom of trellis for stability

3 Crocks in base to ensure good drainage

A Trellis 'Fan' for a Climber or Wall Shrub

A TALL, free-standing container plant adds instant height to a group of pots, creating an attractive display. While a tripod or obelisk is ideal for some climbers, other climbers and wall shrubs need more room to spread out, and this can be easily provided by using a fan-shaped piece of trellis.

First, choose a suitable container. It must be large – preferably a minimum of 45cm (18in) across – and made of a heavy material such as wood, terracotta or stone to provide sufficient weight at the base to prevent a top-heavy plant from toppling over. Place the trellis so that its base is resting on the bottom of the pot. Secure it by tying a couple of pieces of wire around the bottom of the trellis, threading it through the drainage holes and winding it around a large nail or length of metal laid horizontally on the outside.

Put some crocks or stones in the base to ensure good drainage, then add a couple of bricks for extra stability. Partly fill the container with potting compost – a soil-based type is best for permanent plants – put in the plant and fill around it with compost, leaving a gap of about 2.5cm (1in) between the top of the soil and the rim of the pot.

Suitable plants include hardy climbers such as *Jasminum officinale* 'Fiona Sunrise' (syn. *J. o.* 'Frojas'; common white jasmine), *Akebia quinata* (chocolate vine) and *Trachelospermum* species; and annual climbers such as *Lathyrus odoratus* (sweet pea) and *Ipomoea alba* (moonflower). Shrubs such as *Choisya ternata* 'Sundance' (Mexican orange blossom) and *Myrtus communis* (common myrtle) can be tied against a trellis and outward-growing shoots pruned out in spring. Many conservatory climbers can be grown in this way, too.

Late-blooming Beauties – Flowering Plants

With a bit of encouragement, many annuals and tender perennials will keep producing their fragrant blooms for most, if not all of the autumn. Good long-lasting annuals include *Malcolmia maritima* (Virginian stock), *Reseda odorata* (mignonette), *Nicotiana* (tobacco plant) and *Ipomoea alba* (moonflower). Late sowings of seed will be paying dividends now, usually continuing to bloom well after plants from earlier sowings have been consigned to the compost heap. With all seasonal plants, regular dead-heading prompts them to keep flowering – if they are allowed to set seed, plants often feel that their job is done for the year. Several applications of a high-nitrogen liquid feed will work wonders too, often boosting tired plants into a last-ditch spurt of growth.

A number of permanent summer-flowering plants will keep a good show going from summer until well into autumn. Top performers include the evergreen climber *Trachelospermum*, with white or creamy-yellow jasmine-like flowers that have a delicious scent, and *Myrtus communis* (myrtle), which has the double bonus of flowers and aromatic foliage. Many of the roses listed in the Plant Directory (see page 120) flower over a long period, sometimes even until the first frosts.

Two less well-known clematis species, *Clematis flammula* and *C. rehderiana*, wait until autumn for their season of glory. Don't expect the plate-sized blooms produced by so many of the colourful large-flowered hybrids, but instead delight in the profusion of small flowers – white, star-shaped and strongly scented in the case of *C. flammula*; pale yellow, tubular and cowslip-scented for *C. rehderiana*.

Above and right: Two enchanting late-blooming clematis have the bonus of fragrance. *Clematis flammula* forms an absolute mass of white starry flowers and the dainty, bell-shaped pale yellow blooms of *Clematis rehderiana* are cowslip-scented.

A Pot of Layered Bulbs

Bulbs are incredibly versatile, and where space is really limited it's possible to get three times the amount of colour from a single pot by planting layers of different bulbs.The lowest bulbs work their way around those that are higher up to create a spectacular display of colour. The largest bulbs, such as hyacinths or tulips, are planted first at a depth of about 15cm (6in). Next comes a layer of medium-sized bulbs, such as the taller narcissi. Finally come the tiniest bulbs, such as crocus or the smallest narcissi. If you are not sure about planting depths, as a rule of thumb a bulb should be covered with compost to three times its own height.

Good drainage is the key to success with bulbs in pots. Make sure the container has a drainage hole and put a layer of crocks or stones in the base. Support the pot so that it is a centimetre or so off the ground to allow surplus water can drain away easily. In cold areas, keep the pot in an unheated greenhouse or cold frame for the worst of the winter weather or stand it outside in a sheltered spot.

1 *Tulipa* 'Generaal de Wet'
2 *Narcissus* 'Cheerfulness'
3 *Crocus chrysanthus* 'Snow Bunting'

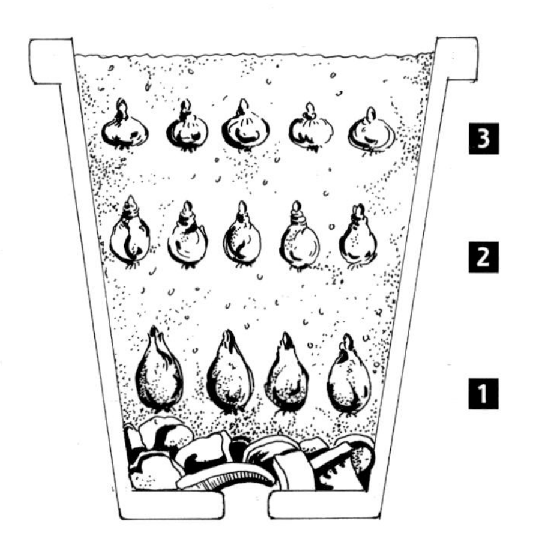

A Succession of Sumptuous Lilies

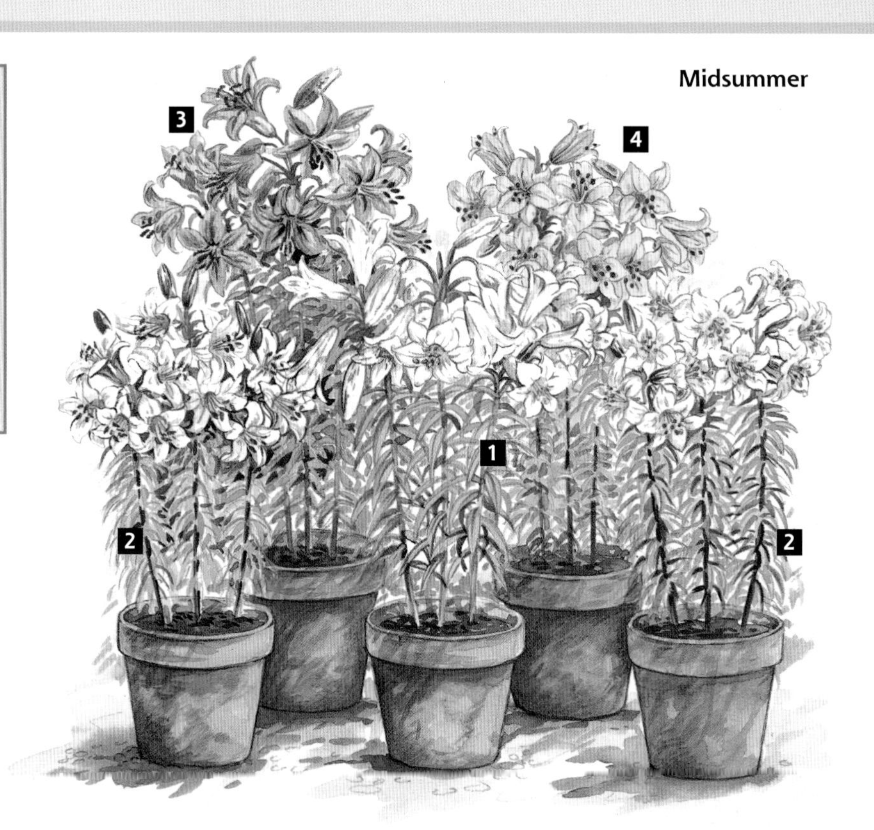

Few flowers create a greater impression of full-blown luxury than lilies, perhaps because they are an expensive cut flower or maybe because their huge blooms give off a rich and exotic perfume that is one of the strongest flower fragrances of all. Lilies are wonderful in pots, and they can be stood close to doors and windows to flood the house as well as the patio with fragrance. For a really sublime experience, move a pot of lilies into a conservatory or garden room to pack it with perfume.

Buy lily bulbs in either autumn or spring and check that they are firm to the touch and free from mould. Plant three bulbs to a deep pot measuring 25–30cm (10–12in) across, placing crocks in the base and planting so that there is 10–15cm (4–6in) of compost over the bulbs. Use a soil-based potting compost. An unheated greenhouse is an excellent growing environment; alternatively, stand the pot in a sheltered spot outside.

Most lilies do best among other containers because the majority like to have cool roots but their 'heads' in the sun. Make the overall effect more decorative by sowing seed of a hardy annual such as *Nigella damascena* (love-in-a-mist) around the base of the lilies in early spring so that their tall stems rise from a low mound of ferny leaves and starry, but unscented, blue flowers.

Choose a selection of varieties that bloom at different times in order to have these gorgeous blooms all summer long.

In midsummer *Lilium longiflorum* (Easter lily, Bermuda lily) has pure white, trumpet-shaped flowers. *L. regale* (regal lily) is the easiest lily of all to grow, and it has a superb perfume. The trumpet-shaped white flowers have a yellow throat and are flushed with purple outside. *L. r.* 'Album' is pure white with a yellow throat. Of the trumpet lilies, *L.* African Queen Group has large apricot flowers, while the flowers of *L.* Golden Splendour Group are deep golden-yellow.

Many oriental hybrids, which bloom in mid- to late summer, have wonderfully perfumed and showy flowers. They need to have their roots in the shade but their heads in the sun and a plentiful supply of moisture. *L.* 'Kyoto' is a tall variety with pure white flowers. *L.* 'Star Gazer' is of medium height; the flowers are deep crimson with a white edge. *L.* 'Mona Lisa' is a compact variety with soft pink flowers. Around these taller lilies stand a couple of pots of *L. formosanum* var. *pricei*, a dwarf lily that bears pure white, trumpet-shaped blooms on short, stout stems.

Troubleshooting

Slugs and snails adore lilies, and the emerging shoots are particularly vulnerable. Place 'collars' made from clear plastic bottles around new shoots until they are well established. Use an environmentally friendly slug and snail killer such as those based on aluminium sulphate. A line of non-setting insect glue or petroleum jelly around the pot's rim will help keep these pests away too.

Lily beetle is becoming a pest in some areas. Watch out for blobs of black slime on the plant's stems and leaves. Inside is an orange-brown larva that will develop into a bright red beetle. Hand-picking is the most effective control as these pests hatch over a long period.

Midsummer

1 *Lilium longiflorum*

2 *L. regale*

3 *L.* African Queen Group

4 *L.* Golden Splendour Group

Late summer

5 *L.* 'Kyoto'

6 *L.* 'Mona Lisa'

7 *L.* 'Star Gazer'

8 *L. formosanum* var. *pricei*

Making the Most of an Unheated Greenhouse

A greenhouse is a wonderful aid to gardening in cold areas, but those without heating tend to be left virtually empty during the winter months. There are, however, lots of plants that can still be grown here from autumn to spring to extend the flowering period for next year. For a bit of extra frost protection, insulate the structure with bubble polythene.

- Sow hardy annuals in autumn, in small pots or modular trays. The plants will grow slowly through the winter to flower in late spring to early summer next year. Make a second sowing, including sweet peas, sweet alyssum, sweet sultan, mignonette and blue woodruff, in early spring for later flowers.

- Container-grown herbs moved under cover can be harvested for a longer period than if left outside.

- Pots of bulbs can be planted up and kept here for the winter, to avoid danger of frost damage.

- Tender perennials can often tolerate the cold if they are kept under cover with their compost barely moist. Do be prepared for losses, however!

- Lift and pot up clumps of bulbs and plants such as snowdrops, primrose and lily-of-the-valley. The protected environment will bring the flowers on slightly earlier. They can also be taken indoors to a cool room if desired.

Seasonal Reminders

Propagation

- Finish taking cuttings of tender perennials to overwinter on windowsills indoors.
- Divide established clumps of herbaceous perennials such as catmint, ground ivy, primula and lily-of-the-valley.
- *Trachelospermum* spp. can be propagated by layering.

Planting

- Biennials such as wallflowers and sweet williams can be bought and planted, or transplanted from a nursery bed, into containers.
- Freesia corms for spring flowers can be planted in a greenhouse or conservatory.
- Spring bulbs should be planted as soon as possible. The exception is tulips, which can be planted later in the season.
- Winter-flowering shrubs are best potted up now to give them time to settle into their containers.

General

- Check that greenhouse and conservatory heaters are working properly. Clean the building thoroughly, both inside and out, and insulate with bubble polythene before it is filled with plants.

Star jasmine (*Trachelospermum jasminoides*) produces its deliciously fragrant blooms through summer and autumn.

- Reduce watering as growth slows down.
- Pots that are not frost-proof should be emptied, cleaned and stored.
- In cold areas, frost-tender shrubs and perennials should be brought under cover before the first frosts.
- *Cosmos atrosanguineus* (chocolate cosmos) and *Mirabilis jalapa* (marvel of Peru) are frost tender but have tuberous roots that can be stored over winter. Dig plants up carefully, cut the stems to leave just a couple of centimetres at the base, put the tubers in trays and cover with old compost so that the crowns are just visible. Keep the soil barely moist and store in a frost-free place.
- Pots of lilies can be brought under cover if space permits. Alternatively, lay pots on their sides in a sheltered spot outside to avoid them becoming waterlogged.

Purple fennel (*Foeniculum vulgare* 'Purpureum'), shown in the centre, is a herb too often scorned for its prolific self-seeding. It looks truly magnificent in a large container where its feathery foliage is an alluring contrast to nearby plants.

Winter – Scarce but Treasured Scents

MANY people consider winter to be a time when the garden is completely bleak and empty, but fortunately there are several shrubs that choose to bloom at this darkest time of the year when our spirits are sorely in need of a boost. These few fragrant gems need to be carefully placed in order to enjoy their scent, as this is rarely a time of year to linger outside. Put winter-scented plants by frequently used doors, gates and pathways and, if you have a conservatory or even an unheated porch, have a couple of extra pots to bring inside and fill the room with perfume. Then, just when it feels as if winter will never end, up come the earliest bulbs, defying the harsh weather to produce delicately scented blooms that bring the promise of spring.

Hardy Winter Winners – Shrubs

Although there are very few evergreens with scented flowers, how lucky it is that a couple of the most valuable ones choose to bloom in winter when fragrance is in very short supply. An essential plant in my view is *Sarcococca* (Christmas box, sweet box), which is a neat, well-behaved and handsome shrub. It forms a compact, upright clump of stems clad with pointed, glossy, dark green leaves, is happy in sun or shade, and in midwinter produces many tassels of creamy-white flowers that are insignificant in appearance but have a powerful sweet scent.

In winter skimmias bear scented white flowers in large, showy clusters, on a rounded bush of oval, dark green leaves. Female varieties produce glowing red berries, but only if there is a male is nearby.

Above: Christmas box (*Sarcococca*) is a paragon. Scented winter flowers are coupled with evergreen foliage, a neat habit and tolerance of sun or shade.

Right: Pink *Cyclamen persicum* shows off well against coloured foliage in this hanging basket. The fragrant blooms last well if protected from frost.

Opposite: Showy red flower buds of *Skimmia japonica* 'Rubella' last all winter and open to scented white flowers in spring. The container is tucked against *Viburnum bodnantense* 'Dawn', a shrub with fragrant pink blooms, which could be grown in a pot for two or three years.

A Group of Pots for Winter

1. *Viburnum* × *bodnantense* 'Dawn'
2. *Sarcococca hookeriana* var. *digyna*
3. *Lonicera* × *purpusii*
4. *Skimmia japonica* 'Rubella'
5. *Skimmia japonica* subsp. *reevesiana*
6. *Hedera helix* 'Glacier' (unscented)
7. *Galanthus nivalis* 'Flore Pleno'
8. *Crocus chrysanthus* 'Snow Bunting'
9. *Crocus chrysanthus* 'Cream Beauty'

A FEW wonderful plants brave the harsh winter weather to bring fragrance and colour to our gardens. Some are deciduous and produce flowers all along their naked branches. *Viburnum* × *bodnantense* 'Dawn' is a large shrub, which bears clusters of pink, sweetly scented on its leafless branches. *Lonicera* × *purpusii* (winter-flowering honeysuckle) is a twiggy shrub, with many small, creamy-white flowers that are very sweetly scented.

The evergreen foliage of *Sarcococca* (Christmas box, sweet box) and skimmias provides welcome structure as well as scented flowers. *Sarcococca hookeriana* var. *digyna* is a compact shrub, with dark green, glossy, pointed leaves and tassels of creamy-white, strongly scented flowers.

Skimmia japonica 'Rubella' is a compact shrub with rounded, purple-edged leaves. Clusters of showy, red-purple flower buds are borne right through winter and then open to reveal fragrant white flowers. *Skimmia japonica* subsp. *reevesiana* is a dwarf shrub that produces red berries in autumn, and these often last all winter. Decorative flower buds open in late winter to early spring. Skimmias prefer to be grown in ericaceous (lime-free) compost. Most varieties are either male or female, and if the female plants are to bear berries, both must be grown together. *S. japonica* subsp. *reevesiana* is, however, a hermaphrodite form and produces both male and female flowers on the same plant. *S. j.* 'Rubella' is male and does not have berries.

The earliest bulbs push up through the soil to produce delicately scented blooms. *Galanthus nivalis* 'Flore Pleno' (double snowdrop) is a popular bulb. The white flowers, marked green inside, have a honey scent. The scented crocuses bloom in late winter; *Crocus chrysanthus* 'Snow Bunting' has white flowers, while those of C. c. 'Cream Beauty' are creamy-yellow.

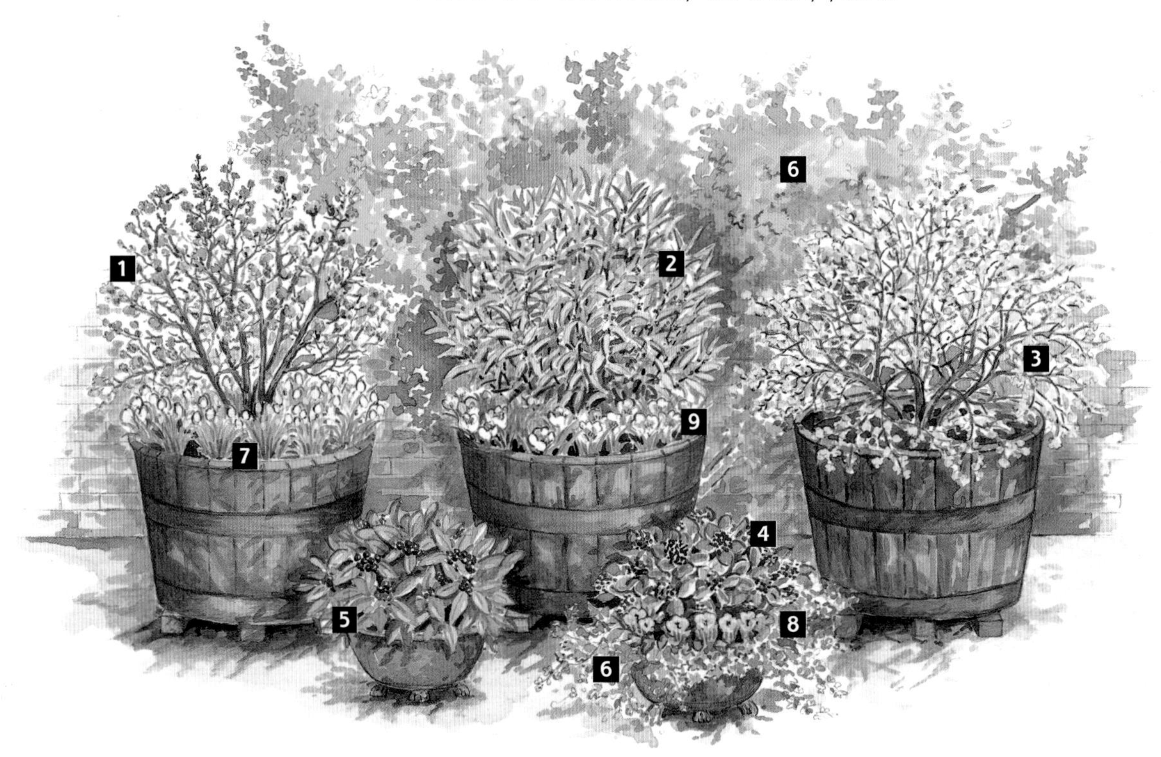

Most compact of all is *Skimmia japonica* subsp. *reevesiana*, which has the great bonus of bearing male and female flowers on the same plant, so it also bears red berries that appear in autumn and often last until spring. Most decorative of the other forms is *S. japonica* 'Rubella', with red-tinged leaves and attractive red flower buds that look good all through winter before opening late in the season.

Mahonia aquifolium 'Smaragd' is a handsome and tough evergreen with glossy but prickly, dark green leaves and clusters of yellow, sweetly scented flowers that open in late winter to early spring.

In addition to the evergreens described above, which are all compact and ideal for containers, there could be added a couple of larger deciduous shrubs. Strictly speaking, these are on the borderline of container growing because after several years they'll need to be planted out in the open ground, but they are well worth considering on a shorter-term basis if pots are your only means of gardening. Look for *Viburnum* × *bodnantense* 'Dawn', an upright shrub that bears clusters of deliciously scented pink flowers on its naked stems in midwinter, and the winter-flowering honeysuckles *Lonicera fragrantissima* and *L.* × *purpusii*, which bear small, white to cream blooms that have a piercingly sweet scent.

Early-blooming Charmers – The First Bulbs

Bulbs are generally thought of in conjunction with spring, but there is a good selection of delightful dwarf varieties that start to come into bloom as early as midwinter. At such a time, it is wonderfully uplifting to see these apparently fragile flowers surviving the often harsh weather to give out their delicate scent. Growing these early-flowering bulbs in pots offers the chance to appreciate them to the full, particularly as most of them have small or downturned blooms. Raise the pots off the ground on a hanging shelf or low wall to bring these shy beauties nearer the nose – after all, it is a trifle difficult to grovel full-length on the ground to appreciate the fragile tracery of green inside a snowdrop flower and to sniff its honey scent.

Galanthus spp. (snowdrops) are first to flower, and although their soft fragrance certainly pales beside the powerful perfumes of later blooms such as hyacinths, the very fact that they flower at all at such a gloomy time makes them immensely appealing. The double form of the common snowdrop is more scented than the single; stronger still is the perfume of *G.* 'S. Arnott', a large and vigorous variety. In keeping with the delicate simplicity of the flower, I like to grow clumps of snowdrops individually in small, plain, terracotta pots, with a little moss to cover the soil. These can be ranged along a low wall or on a small shelf at shoulder height, and if the weather is too horrid to linger outside, bring the pots into a cool room or porch so that you can appreciate them in comfort.

Plants to Protect in Cold Areas

Choisya ternata 'Sundance' brings its own winter sunlight to the garden but needs protection.

Certain plants that are on the borderline of hardiness will generally need winter protection in cold areas. These include:

Akebia (chocolate vine)
Choisya ternata (Mexican orange blossom)
Daphne odora
Laurus nobilis (bay)
Lavandula (lavender)
Myrtus communis (myrtle)
Pieris
Rosmarinus (rosemary)
Thymus (thyme)
Trachelospermum

Ideally, move these under cover into an unheated greenhouse, porch or conservatory. If none of these options is available, protect them as described on page 101.

In the depths of winter the first bulbs lift the spirits beyond measure. Snowdrops (*Galanthus*) and *Iris reticulata* have a delicate scent too.

Pots of Spring Jonquils

Of all the narcissi, it is the jonquils that, in my opinion, have the most outstanding perfume. All are excellent in pots, growing from 15cm to 45cm (6–18in) high and bearing small, cupped flowers, usually in small clusters but sometimes singly. A group of jonquil narcissus and its hybrids, with separate varieties in five individual pots so that their heights contrast effectively, makes a charming display with a uniting overall theme.

Choose deep, plain terracotta pots. The two larger ones should be 25–30cm (10–12in) high and 23cm (9in) wide at the top. To make a contrast of shape, use slightly smaller half-pots for the dwarf varieties – pots about 20cm (8in) high and 17.5cm (7in) across would be ideal. Put crocks in the base of each pot to ensure good drainage and use a soil-based potting compost. Use 10 bulbs of each variety in each pot. Plant in early autumn and keep in a sheltered spot in winter, ideally in an unheated greenhouse or cold frame, and move outside in spring.

Narcissus jonquilla (often called the single jonquil) has single, lemon-yellow flowers that have a simple and uncomplicated charm; it grows to 30cm (12in). *N.* 'Baby Moon' has golden-yellow flowers and grows to 20cm (8in). The flowers of *N.* 'Pipit' are soft lemon, with cups that quickly turn to white; it grows to 23cm (9in). *N.* 'Suzy' has wide, golden petals and shallow orange cups; it grows to 50cm (20in). *N.* Trevithian' is lemon-yellow and grows to 45cm (18in).

After flowering has finished, remove the dead flowerheads. Keep the compost moist and apply a high-potash liquid feed once a fortnight until the leaves begin to yellow. Stand the pots in a sun-baked spot for summer and move under cover again for winter.

1 *Narcissus jonquilla*
2 *N.* 'Baby Moon'
3 *N.* 'Pipit'
4 *N.* 'Suzy'
5 *N.* 'Trevithian'

Forcing Lily-of-the-valley for Winter Flowers

THE technique for 'forcing' – that is, bringing plants into early flower for indoor use – was practised widely by the Victorians but is largely neglected today. *Convallaria majalis* (lily-of-the-valley) is ideal for forcing, and its stems of waxy, bell-shaped, white flowers have a delectable perfume that can scent a whole room. And if you already have clumps growing in your garden, it won't cost a penny!

In midwinter dig up healthy clumps of the plant, shake off excess soil and put the roots in polythene bags. Place them in the refrigerator for two or three weeks. Then pot them up into an ornamental container, using a general-purpose potting compost. Water well and put on a warm windowsill, keeping a minimum temperature of 18°C (64°F) during the day and 13°C (55°F) at night. Once the flower buds can be seen, move the pot to a cool room with a minimum temperature of 5°C (41°F). Feed weekly with a liquid fertilizer from the time when the leaves are well formed until the flower spikes have developed.

After flowering, replant outside and leave the clump for several years to recover before forcing again. If you become addicted to this gorgeous plant, it's a good idea to have a nursery bed outside with a row of plants ready for lifting and forcing.

Simple pots of lily-of-the-valley (*Convallaria majalis*) are placed on a shelf so their seductive perfume can be appreciated to the full. They are also a perfect example of how several small pots look better than a single larger one.

Blooming in late winter, *Crocus chrysanthus* 'Cream Beauty' has a sweet honey fragrance.

Iris reticulata is another good, dwarf early-flowering bulb that can be treated in the same way; the rich purple flowers are beautifully marked with orange-yellow and smell of violets. Snowdrops are woodlanders, preferring a site in dappled shade, while the iris needs a sheltered site in full sun, together with sharply drained compost.

The early crocus species and varieties offer some delightful flower colours, with a lovely honey scent that is best appreciated if containers are raised off the ground towards nose height. *Crocus chrysanthus* is the most valuable scented type, and it blooms in late winter. Named varieties include *C. c.* 'Cream Beauty' (pale creamy-yellow), *C. c.* 'E.A. Bowles' (yellow) and *C. c.* 'Snow Bunting' (white).

Caring for Outside Containers in Winter

In cold areas, plants in pots are very susceptible to frost damage because the entire plant, roots and all, is above ground. Here are a few tips for successful winter pots.

- Good drainage is vital because a sodden rootball is likely to freeze solid. When planting, ensure pots have drainage holes and put a layer of crocks at least 5cm (2in) thick in the base. Raise containers a little way off the ground so that surplus water can drain away.
- Insulate containers with a layer of polystyrene or thick layers of newspaper, placed around the inside (but not the base) of the pot when planting.
- In very cold spells, move plants close together in a sheltered spot, such as against a wall. Wrap the pots in bubble polythene. The foliage of evergreens can be wrapped in horticultural fleece to protect the leaves from frost and wind scorch.
- The smaller the container, the more susceptible it is to the cold. If possible, put small pots in a greenhouse, porch or cold frame in really cold spells.
- Water very sparingly if the compost is drying out.

Seasonal Reminders

Propagation

- Akebia and jasmine can be propagated by layering.
- Roses can be propagated by hardwood cuttings.
- Continue sowing hardy annuals to bloom early next summer.

Planting

- Plant spring bulbs as soon as possible. Tulips, however, can be planted until almost the middle of winter.
- In late winter, buy and plant snowdrops just after they have flowered (commonly referred to as 'in the green') because they establish best at this time.

Pruning

- Wisteria should be winter-pruned (see page 59).
- Cut back dead foliage of perennials in late winter.

General

- Scrub out used seed trays, pots and tubs in readiness for planting. Old compost can harbour pests and diseases.
- Lift and compost annual plants. Remove any diseased plant material and put it in the dustbin.
- In the greenhouse good hygiene and air movement go a long way towards preventing fungal diseases. Space plants well apart, ventilate whenever the weather permits and remove any dead leaves or flowers as soon as possible. Wash the glass to make the most of scarce winter light.
- Order seed catalogues so that you can fill the dark evenings with dreams of next summer's glory.

Inside-outside Plants – From Conservatory to Patio

In cold areas a heated structure, like a greenhouse or conservatory, makes it possible to grow some delectably perfumed plants such as citrus. All these frost-tender charmers can go out on to the patio for the summer.

In cold areas a conservatory or greenhouse is a fantastic boon to growing a considerable range of wonderfully scented plants that need a frost-free environment for the winter. However, this doesn't mean they need to stay under cover right through the year. Once all danger of frost is past, these tender plants can be moved out onto the patio to provide a ready-made display that has a tremendously exotic flavour. In mild areas, many of these plants will be able to remain outside all year round.

The Warm Conservatory

A heated conservatory can house a wide selection of fragrant plants. An ideal minimum temperature for the winter is around 10°C (50°F), which will make a pleasant environment for yourself as well as for your plants. Many people ask what is the lowest temperature at which plants could survive – it is around 5°C (41°F), although this is a bit on the chilly side and many winter-flowering plants are reluctant to bloom at such low temperatures.

The plants discussed in this chapter are just a selection of some of the best fragrant varieties.

A Lemon-scented Theme

A GROUP of plants can be united by a particular theme, such as those with a citrus scent. All the plants chosen for this display have foliage that, when bruised or crushed, gives off a strong and attractive lemon fragrance. In cold areas nearly all require the protection of a frost-free greenhouse or conservatory for the winter, but they can go outside on the patio in summer.

A lemon tree is the obvious choice for a centrepiece. Lemons and other citrus are surprisingly straightforward to grow (see page 91), and they benefit greatly from spending the summer outside on the patio. *Citrus × meyeri* 'Meyer' is the best variety for planting in a container. It has a compact habit and glossy, evergreen leaves, and it freely produces star-shaped white flowers in clusters over a long period. Fruit will be produced if the temperature is sufficiently high.

Eucalyptus citriodora (lemon-scented gum), which grows to 1.5–1.8m (5–6ft) high, has tall stems clad with long leaves, making a good contrast to the rounded lemon tree. Lanky plants can be cut hard back in spring, almost to ground level.

Aloysia triphylla (syn. *Lippia citriodora*; lemon verbena) is a deciduous shrub that can be grown outside in mild areas if it has the protection of a sunny wall. Slender, lance-shaped leaves, which are slightly ridged and a fresh light green colour, cover the twiggy stems, and open panicles of tiny, lilac-white flowers are borne in summer. Lemon verbena can be grown as a bush, but it's also very effective when it is trained as a standard in a similar way to wisteria (see page 59). A refreshing lemon tea can be made by infusing a few leaves in boiling water for several minutes.

Several of the scented-leaved pelargoniums (geraniums) have a delightful lemon scent, including *Pelargonium crispum* 'Variegatum', *P.* 'Graveolens', *P.* 'Mabel Grey' and *P.* 'Lady Plymouth'. These also need a frost-free environment, although they are small and decorative enough to be kept on a windowsill.

Lastly, there are two easily grown herbs that are frost-hardy in a sheltered site – *Thymus × citriodorus* 'Aureus' (lemon thyme) and *Melissa officinalis* 'Aurea' (lemon balm). For maximum ornamental value, I have chosen these two varieties because they have gold-variegated leaves. Grow them in shallow containers and trim in summer to maintain a neat shape.

1 *Citrus × meyeri* 'Meyer'
2 *Eucalyptus citriodora*
3 *Aloysia triphylla*
4 *Pelargonium* 'Mabel Grey'
5 *Pelargonium* 'Lady Plymouth'
6 *Thymus × citriodorus* 'Aureus'
7 *Melissa officinalis* 'Aurea'

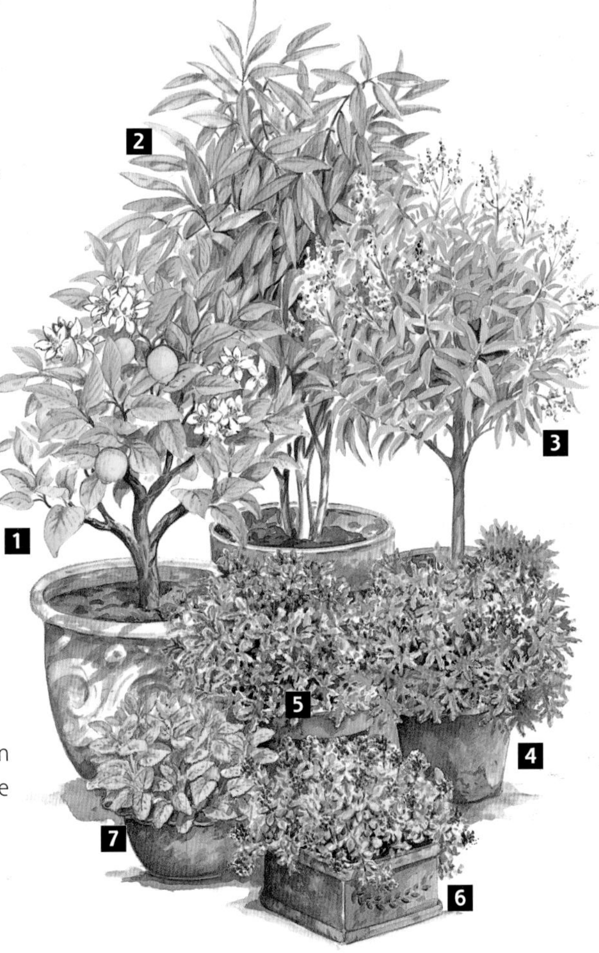

Brugsmania (angel's trumpets) produces a spectacular show of exotic and richly scented blooms, but do bear in mind that all parts of the plant are very poisonous.

Winter is the time when a conservatory is most appreciated and there are some superb plants that lift one's spirits in these gloomy days. Many plants are, in any event, evergreen, so there is a plentiful backdrop of foliage. In addition, a number of varieties produce wonderfully fragrant blooms at this time. Two tender buddleia species – *Buddleja asiatica* and *B. auriculata* – are evergreen, with, respectively, white and cream flowers, which are powerfully scented. Of the climbers, *Jasminum polyanthum* is a familiar pot plant, which will romp away if given a large pot, bearing masses of white flowers that have a strong, almost overpowering scent. Any small spaces can be filled with pots of *Primula kewensis*, which bears golden-yellow, sweetly scented blooms from winter into spring.

Once spring gets into its stride, plenty more plants begin to bloom. *Boronia megastigma* is an excellent small shrub with lemon-scented flowers and aromatic foliage. A lemon tree adds a touch of luxury, and the best compact variety, *Citrus* × *meyeri* 'Meyer', flowers virtually all year round (see page 91). For architectural effect try the marvellous *Euphorbia mellifera* (syn. *E. longifolia*; honey spurge), which has long, slender leaves and clusters of yellowish flowers that have an outstanding honey scent, or *Pittosporum tobira* (Japanese pittosporum, Japanese mock orange), a superb evergreen with clusters of creamy-white, heavily scented flowers that is best grown as a standard (with a clear stem and a rounded head) where space is limited.

Summer-flowering shrubs include the spectacular angel's trumpets or *Brugmansia* (formerly *Datura*), a large shrub bearing splendid, dangling trumpets of heavily scented flowers – though bear in mind it is very poisonous. Unless space is particularly plentiful, grow the angel's trumpets as a small standard tree and prune back to the top of the stem each year after flowering. Climbing plants can be grown on a trellis pyramid or fan (see pages 49 and 71). Good candidates for such treatment include *Mandevilla laxa* (syn. *M. suaveolens*; Chilean jasmine), which has very large, white, jasmine-like blooms; *Jasminum azoricum*, which bears beautiful flowers right through summer and autumn with a perfume that is less overpowering than the winter-flowering *J. polyanthum* (see above); and *Dregea sinensis* (syn. *Wattakaka sinensis*), which bears honey-scented flowers that are white streaked with pink.

The arrival of autumn in the conservatory simply signifies a range of different plants coming into bloom, and these are a wonderful antidote to the slow dying of the year outside. Some are very unexpected – camellias, for example, are generally associated with spring, but *Camellia sasanqua* 'Narumigata' bears fragrant autumn flowers that are white tinged with pink. *Eriobotrya japonica* (loquat) starts to bloom in late autumn to early winter, bearing panicles of creamy-yellow flowers that have a spicy, apricot fragrance. A dramatic architectural plant, with dark glossy green leaves that can be up to 30cm (12in) long, loquat makes a superb tall specimen. Freesia bulbs can be planted now to bloom in winter or spring.

Caring for Citrus

The sheer luxury of harvesting your very own lemons and other citrus fruit need not be an unattainable dream. Citrus trees are, in fact, quite amenable to container cultivation and will tolerate surprisingly low temperatures. Here are a few guidelines.

- Pot young plants into 30cm (12in) terracotta pots, using a soil-based potting compost with a little extra grit to improve the drainage.
- Citrus plants prefer to be outside in summer. Acclimatize them gradually by putting them outside in light shade for two or three weeks before moving them to a sunny spot. Do the same but in reverse when bringing them in for winter.
- Stand the pot on a tray of wet gravel to ensure reasonable humidity around the plant.
- In spring and summer feed once a fortnight with a tomato fertilizer. If the leaves begin to yellow, feed with sequestered iron.
- Scale insect is a common problem and can be controlled by rubbing off the scales with a cloth soaked in methylated spirit.
- Keep the tree in shape by pinching out the growing tips once the branches have reached the desired length.

Few plants appear so exotic as citrus, but they are surprisingly straightforward to grow.

Moving Plants Around

Shifting well-established plants is no mean task, and care must be taken to avoid pulling muscles or wrenching backs. A 'sack-barrow' trolley is a very worthwhile investment for moving heavy pots as well as other items. Although one person can then move large plants with reasonable ease, a second person is invaluable to stop plants toppling over and to help site them in just the right place. A very large plant can be moved on its own little wooden trolley with castors attached. Fix a large hook on each end of the trolley, so that ropes can be attached when moving becomes necessary. Lightweight plastic pots help cut down on weight, although do remember that plants tend to grow better in thicker pots where there is less fluctuation in temperature around the roots.

Scented foliage can easily be overlooked in favour of flowers, but I recommend including at least a couple of delightful aromatic plants such as *Salvia elegans* 'Scarlet Pineapple' (syn. *S. rutilans*; pineapple sage), *Prostanthera cuneata* (alpine mint bush) and *Aloysia triphylla* (lemon verbena), which have the delicious scents that their common names suggest.

The Unheated Conservatory

Although a heated conservatory offers by far the greatest scope, a surprising number of tender plants can survive if they are overwintered in an unheated structure. Evergreens particularly will benefit by being protected from freezing winds, which often cause unsightly leaf scorching. The key to success is to keep the compost barely moist so that if hard frosts do occur, the rootball is far less likely to freeze and cause root damage.

Hardy plants also have tremendous potential for use in an unheated conservatory. Certain varieties will really benefit from its protection in cold areas (see page 81) and their fragrance can be appreciated to a far greater degree here, where one can actually sit and take it in, as opposed to outside, where conditions are rarely good enough to linger. All the winter-flowering plants on pages 78–83 can be used in this way.

Conservatory Care

Many conservatories seem to be constructed with the aim of simply adding an extra room to the house rather than for growing plants, and this has the consequence of creating an environment that is positively hostile to plants. A structure that is poorly ventilated and exposed to full sun is far from ideal; do bear in mind that plants need plenty of ventilation and at least some shade in order to thrive indoors in summer. Of course, this is not a problem if the plants are just overwintering under cover and spending the summer outside. Here are some basic care tips.

Watering and Feeding

Both under- and overwatering are traps for the unwary! Too much water is a common cause of problems during winter in particular, a time when growth has slowed right down and plants should be allowed to become almost dry between waterings. In contrast, a plentiful and regular supply of water is needed on warm summer days, when plants can dry right out in hours. An automatic watering system can be an excellent investment (see page 101). Feeding is necessary from spring to early autumn, but only during winter if the plant is in a warm environment and growing actively.

Lemon verbena (*Aloysia triphylla*) has leaves that are delightfully aromatic when crushed. This 'must-have' plant can be grown as a standard, a bush or trained against a warm wall.

Conservatory Calendar

Spring

- Reduce daytime heating and increase ventilation as the outside temperature rises.
- Increase watering as growth begins in earnest.
- Many plants need pruning now, particularly if they are several years old.
- Top-dress plants with a little fresh compost and controlled-release fertilizer.
- Repot potbound plants, moving up one size of container.
- Control and monitor pests using yellow sticky traps.

Summer

- Put up shading.
- Ventilate as much as possible.
- Move the plants outside if desired, once all danger of frost is past.
- Water frequently and spray paths with water to increase humidity on hot days.
- Introduce a biological control if any pests have become a problem.
- Most plants can be propagated by half-ripe cuttings.

Autumn

- Empty the conservatory and clean it thoroughly.
- Check that heaters are working properly.
- Move tender plants back under cover before the first frosts.
- Reduce watering as growth slows down.
- Stop feeding plants, except for those that are flowering.

Winter

- Clean the glass inside and out to make the most of available light.
- Water sparingly, letting the compost almost dry out between waterings.
- Prevent disease by spacing plants well apart to allow good air movement.
- Ventilate whenever weather conditions permit.
- Regularly remove any dead leaves or flowers, which can harbour pests and diseases.

Ventilation and Shading

Plenty of fresh air is essential for healthy growth, and too much heat in summer can kill plants. Make sure there are vents in the side as well as in the roof, and it's also worth installing an automatic vent opener. If there isn't a roof vent, install an extractor fan.

Unless the structure is in a shady site, some degree of shading is essential to filter out strong sunlight. Attractive blinds can be fitted in a conservatory that also does duty as an extra room, while whitewash-type shading can be applied to a more functional building.

Pests and Diseases

A warm, protected environment is, unfortunately, ideal for pests and diseases as well as for plants. Many problems can be avoided in the first place by keeping plants growing strongly, because a plant that is under stress from lack of food or water or from being grown in conditions it dislikes is much more susceptible to attack. Good hygiene also goes a long way towards avoiding problems. Every few days, remove and throw out any dead leaves and flowers that can be the source of disease, and space plants well apart to allow for good air circulation. Ventilate as much as possible in winter as weather conditions permit.

Some pest problems are virtually inevitable in a conservatory or greenhouse, but don't despair. Vigilance is the key to success, so inspect plants frequently for signs of attack – hang up yellow sticky traps, look under the leaves and check by torchlight at night for nocturnal pests so that you catch the infestation in the early stages rather than waiting until it has become widespread. Many pests can now be combated by biological controls (introducing an insect to eat the pest), which is safer and often more effective than using harmful chemical sprays.

Pruning and Training

Many conservatory plants need regular pruning and training. If they are left unchecked, they would quickly becoming sprawling, tangled or just perform poorly. Details are given for individual plants in the Plant Directory (see pages 121–23). Climbers growing on pyramids or trellis fans need very frequent training to weave their shoots round the support.

Practical Matters

PLANTS in containers rely on you for all the essentials of life, and so choosing the right plants and pots is only half the key to success – the remainder depends on how planting is done and on regular care and attention. Starting with the basics of choosing composts and fertilizers, and how to plant, then following through with maintenance details, overwintering, troubleshooting and propagation, this compact guide will provide you with all the necessary information for growing absolutely first-class plants in containers right through the year.

Choosing Compost for Pots

When a plant's roots are confined to a limited growing space, it is essential that it grows in good potting compost if the plant is to perform well and consistently. Skimping on this aspect is a false economy. A wide range of composts is available, but take care not to confuse potting composts, which are specially formulated for container growing, with planting composts and soil conditioners, which are designed for use in garden borders.

The main choice is between soil- or loam-based compost made to the John Innes formula, and soil-less composts. Loam-based compost is heavy, and ideal for pots of permanent plants because it provides more of a buffer against drought than a soil-less mixture, although the pots have to stand on the ground because of the weight of the compost. For this reason, choose soil-less composts for windowboxes, hanging baskets and other hanging containers, where weight is an important consideration. Plants that dislike lime and prefer a neutral or acid soil should be grown in ericaceous (lime-free) compost.

Two exceptions can be made to using 100 per cent fresh potting compost. One is with very large pots, where a 50:50 mix of potting compost and well-rotted garden compost will still give good results. The other is with bulbs, which already contain their own store of energy, and so they can be planted in compost that has previously been used for growing summer-flowering plants or in the soil from old growing-bags. For optimum results, it's best to mix the old soil with an equal amount of fresh potting compost.

Fertilizers

All potting composts contain enough fertilizer to last for about six weeks. After that, it's up to you to provide extra food to keep your plants in top condition during the growing season (there's no need to feed in autumn or winter). Seasonal summer plants are particularly in need of feeding

Young plants, raised from cuttings and in the process of growing on, create an attractive, ornamental effect in this simple wooden shelving unit.

because they perform at full burst for a comparatively short time – don't wait for them to sicken or yellow before beginning to feed.

There are two main options when it comes to feeding. Liquid fertilizer needs to be diluted in water and applied once or twice a week in most cases. For flowering plants, use a high-potash feed, such as tomato fertilizer, because potash boosts the production of both flowers and fruits. For summer-flowering annuals, it's best to switch to a high-nitrogen fertilizer in late summer to encourage a final burst of growth. Don't apply a liquid feed to dry compost or the plant's roots could be scorched.

Controlled-release fertilizer should be added to the compost in spring or early summer. It comes in the form of pellets or small granules coated in a slow-dissolving resin that is temperature-sensitive, so nutrients are released only when the plants are actively growing. The nutrients last for most of the season, although towards the end of summer it's worth topping up annuals and tender perennials only with a weekly liquid feed until the end of the season. Perennial plants shouldn't be fed at this time of year because it would result in soft growth that is susceptible to frost damage.

Preparing and Planting Containers

Pots that have been used previously should be scrubbed out using hot, clean water and a stiff brush because any old debris can harbour pests and diseases. Stone and terracotta containers are porous and are best soaked in water for an hour or so before planting or they'll absorb water from the compost. To reduce water loss from such containers, line the sides (but not the base) of the pot with thick layers of newspaper or a layer of polythene.

Good drainage is essential because roots need air as well as water, and a plant that is sitting in waterlogged compost will soon die. First, make sure that all containers have drainage holes, then put a layer of broken pots (crocks) or pieces of polystyrene in the base of the container to stop the holes becoming blocked. Ideally, cover this drainage layer with newspaper or fine plastic mesh to stop compost clogging up the gaps. After planting, raise the containers a centimetre or so off the ground so that surplus water can flow away. Use ornamental pot feet, stones or old tiles placed underneath the pot.

Thoroughly water all plants an hour or so before starting to plant, since a dry rootball is very hard to re-wet afterwards. Then, put a layer of compost in the container to reach as far as the bottom of the rootball of the largest plant. Knock the plants gently out of their pots by holding each one upside-down with your fingers spread across the surface of the rootball, then tap the pot's edge on a wall or step to loosen it. Place the plants in the container, filling the gaps between them with compost and

Planting a Container

Water plants thoroughly an hour or so before planting.

1 Plant so that rootballs are at the same depth as they were previously growing.
2 Line stone and terracotta pots with newspaper to reduce water loss.
3 Place a layer of crocks, 5cm (2in) deep, in the base to provide good drainage.
4 Make sure the drainage hole is clear.
5 Always use fresh potting compost.
6 Leave about 2.5cm (1in) between the top of the compost and the rim of the pot as space for watering.

firming it gently with your fingers as you go. Don't press too hard – it's easy to over-firm and compact the compost. Finally, use a watering can or a hose fitted with a fine rose to water the container thoroughly, and top up the compost if it settles and leaves gaps. Aim to leave a space of about 2.5cm (1in) between the top of the soil and the rim of the pot so that it can be watered easily.

Planting a Hanging Basket

Set the basket on a bucket or a large pot for stability and stand this on a surface at waist height. Line the basket with a special liner made from wool, coir, foam or card; it is best not to use sphagnum moss, which may have been harvested from threatened wild areas. Be sure to buy the right size liner to match your basket.

Fill about a third of the basket with compost. The secret of successful hanging baskets is to cram in as many plants as possible so, starting at the base, make holes in the liner and put in several trailing plants. Use young plants with small rootballs for planting through the sides and base. Continue to plant and add compost in stages as you work up the basket.

Planting a Hanging Basket

Stand the basket on an empty pot. Line it and fill it about a third full with compost. Start with the smaller trailing plants, gently compacting the rootballs to push them through the mesh.

Continue working up the basket. Leave a gap of about 2.5cm (1in) between the top of the compost and the rim of the basket, for ease of watering. Larger, bushy plants can be placed in the basket's centre.

The plants that go in the top of the basket can be larger, more established ones if you wish because there's no problem with the size of their roots. For a balanced appearance, put the tallest plants in the centre. Firm the compost gently, but take care not to over-compact it.

If the basket is going straight into its final position, hang it up before watering for the first time, because the compost will soak up lots of water and the basket will obviously become much heavier and harder to lift. Use a watering can fitted with a rose to give the basket a thorough and gentle watering to settle the compost around the plants' roots. Top up the compost if it sinks and leaves any gaps.

Watering

Frequent watering is one of the most important aspects of growing plants in containers, because rain is very rarely sufficient. Bear in mind that hanging baskets, wall pots and raised containers in particular will dry out very rapidly indeed and usually need watering twice a day in summer. With all pots, aim to keep the soil consistently moist so it neither dries out nor becomes waterlogged. If possible, water in the early morning and evening, when less water will be lost through evaporation and there is no danger of wet foliage becoming sun-scorched.

Caring for Permanent Plants

Container-grown hardy plants such as perennials, shrubs and climbers benefit from a little attention each spring to keep them at their best. Top-dress the plant by gently scraping off the top few centimetres (about two inches) of soil and replacing it with fresh potting compost mixed with some controlled-release fertilizer.

Plants won't perform at their best once they have become pot-bound and packed their containers with roots. Once this stage is reached, either pot up into a larger container or plant out in the border. Perennials can be rejuvenated by dividing their rootstocks into several pieces, discarding the old, woody centre and replanting the smaller sections.

Winter Care

Good drainage is particularly important in winter because soggy compost can freeze and damage the plant's roots, even killing it in severe cases. So, bear this in mind when planting and make sure the pots have a good layer of drainage material. They should also be raised just off the ground so that surplus water can easily drain away.

In very cold weather the roots of all container-grown plants will be vulnerable to frost damage because the whole rootball is above ground, and it's well worth insulating the pots as a temporary measure when hard frosts are forecast. Stand the pots close together – this alone will give a degree or so increase in temperature – and if possible move them against a wall for extra protection. Wrap the pots with a material such as bubble polythene or hessian stuffed with straw. Evergreen plants benefit from additional protection because the leaves can become scorched by frost or bitterly cold winds. Wrap the whole plant in bubble polythene or horticultural fleece, but take it off as soon as weather conditions improve. Snow is an excellent natural insulator, so let it lie on the plants unless its weight is likely to damage them.

Don't overlook the need for a little watering during the winter months, particularly for containers in sheltered sites, but water them sparingly only if the compost is drying out.

Overwintering Tender Plants

In cold areas frost-tender plants need to be moved under cover into a greenhouse, porch or conservatory, ideally one that is heated to around 5°C (41°F). However, if there is only an unheated structure available, this will be better than nothing, but do bear in mind that survival isn't guaranteed. Insulate the structure with bubble polythene and, most

Watering Systems

A watering system is well worth considering for anyone who isn't at home all the time or even just for holidays. A number of irrigation systems are now available that have been specially designed for the ordinary garden, and the addition of a tap-mounted water-timer will make the system fully automatic.

Irrigation systems may look complex but they are actually quite straightforward to set up. A 'mains' pipe runs from the tap, via a pressure reducer and around the area to be watered. From this pipe come small 'spaghetti' lines to each container, and these lines terminate in little drip heads that deliver a litre (1¾ pints) or so of water an hour. This gentle delivery of water is also much better for the plants than an occasional gush from a hose.

First, position the containers exactly where you want them in order to measure how much hose you will need and the number of containers to be watered. Install your system on a warm day if possible because the pipes will be more flexible. Set out the mains pipe in an unobtrusive position, usually along the base of a wall, and use right-angled joints to fit the pipe snugly around corners. Then, make holes in the pipe for the small individual lines to run to each pot, cutting the line to length afterwards. Fit the drip nozzle into the end of the line; this can be a tight fit, and dipping the line into very hot water will make the plastic more flexible. Finally, peg each line securely into the container. Once your irrigation system is set up, you'll have lots more time to sit back and relax – that is, if you don't get diverted into other jobs in the garden!

Propagating with Half-ripe Cuttings

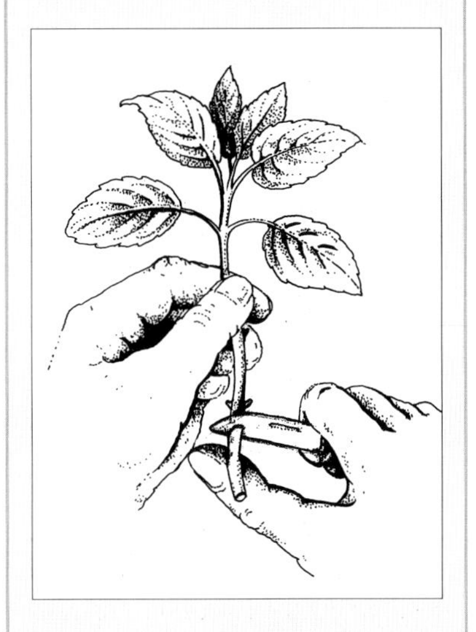

Using a sharp knife, cut just below a leaf joint.

Remove the leaves on the lower two-thirds of the cutting and dibble into a pot of compost. Cover with polythene until rooted.

importantly, keep the compost very much on the dry side since too much moisture is a killer. If there is nowhere at all to bring plants under cover, take cuttings in late summer and keep them over the winter on a windowsill indoors.

Propagation

Many plants can be propagated with reasonable ease, and raising your own plants from scratch can save a considerable amount of money as well as being immensely rewarding. See individual plant details in the Plant Directory (pages 107–23) for the preferred method of propagation.

Cuttings

Cuttings can be divided into three main types for the purposes of the plants described in this book.

Softwood cuttings are taken from the shoot tips in spring or early summer, from growth that is still green and soft.

Half-ripe or semi-ripe cuttings are taken, from midsummer to autumn, from shoots that have just started to become woody. Select healthy, non-flowering shoots and take cuttings 8–10cm (3–4in) long. With a sharp knife, trim the base just below a leaf joint. Take off the leaves on the bottom two-thirds of the cutting. Put the prepared cuttings into seed and cuttings compost mixed with equal parts of perlite or horticultural vermiculite in seed trays or, preferably, into modular trays or small pots, which make it possible to pot on the rooted cuttings with minimum root disturbance. The compost should be kept moist but not over-wet. Cover with a clear plastic cover or polythene, taking care that it doesn't touch the leaves. Take the cover off once the cuttings have rooted, which should be in two to four weeks' time. The following spring, pot up the young plants individually.

Hardwood cuttings can be used to propagate some deciduous shrubs, including roses. In autumn, select healthy shoots about 23cm (9in) long and cut just below a leaf joint. Make a narrow trench in the ground and put a layer of sand in the base to ensure good drainage. Stand the cuttings in the trench so that about two-thirds of their length is below ground and firm in well. Leave for between six months and a year until rooted.

Layering

Shrubs and climbers that are difficult to propagate by other means can often be propagated by layering. Take a young shoot that is near the ground and bend it down so that it touches the soil. It may be necessary to use a small pot filled with compost for this purpose if the parent plant has filled its own pot. Remove a sliver of bark from the underside of the shoot where it touches the soil and peg it down firmly using a 'hairpin' of wire. Leave for up to a year until well rooted, when the layer can be detached, potted up and grown on.

Raising Plants from Seed

Annuals and bedding plants are raised from seed sown from late winter to early spring, usually in pots or seed trays. Biennials are sown outside in early to midsummer, ideally in a nursery bed.

To sow seed, fill a pot or tray with compost and firm it lightly. Using a watering can fitted with a rose, water the compost an hour or so before sowing, because watering afterwards can take the seeds deep into the compost where they won't be able to germinate. Scatter the seed thinly on the surface, then cover if necessary. The seed can be covered with a fine layer of sieved compost, although I prefer to use perlite or horticultural vermiculite, both of which give very good results. This substance is ideal for the few types of plant that need light in order to germinate, as it lets light through while still retaining moisture. Then, cover the pots or trays with glass, clear polythene or plastic film, which also helps to retain moisture. If light is not required, put the trays in a warm, dark place, such as the airing cupboard, and remember to check them every day. As soon as the first signs of germination appear, remove the cover and move the tray to a warm place out of direct sunlight. When the seedlings are large enough to handle, prick them out individually into small pots or line them out in trays. Handle the seedlings only by their leaves, as the stems are easily bruised.

Pricking out Seed

Using a dibber, gently loosen the roots of each seedling. Hold the seedling by the leaves only as stems bruise easily.

Transplant into individual pots. Make a hole for the roots first, then lightly firm in.

Troubleshooting

A few pests, diseases and disorders are fairly common among plants grown in containers and these are outlined below. However, this is not intended to be a comprehensive list of all potential problems. Keeping plants healthy with regular watering and feeding will go a long way towards preventing problems, as healthy plants are much more able to withstand attacks.

Pests

Vine weevil Scalloped holes in the leaf edges are caused by adult beetles which are active at night. The creamy-white, brown-headed larvae live in the soil and feed on the roots, so their presence is often only obvious when the plant begins to collapse. No effective chemical control is currently available, though it is possible to buy potting compost containing an insecticide that will kill the larvae. A biological control is also available, but it is only effective at temperatures above 12°C (55°F). Catch the flightless adult beetles by applying a band of non-setting insect glue around the rim of the pot. Keep the area clear of leaves and debris in which the adults hide during the day.

Slugs and snails Usually less of a problem with container plants than with plants growing in the ground, but they can be troublesome all the same. Active at night, these pests eat holes in leaves and through tender stems. Lilies are a particular favourite of theirs. Put a layer of sharp grit on top of the container and apply a band of non-setting insect glue around the container's rim. Slug pellets are effective, but they contain a persistent chemical; baits based on aluminium sulphate are more environmentally friendly. Encourage natural predators like birds, hedgehogs, frogs and toads, which will do the job for you.

Aphids Clusters of tiny insects appear on young foliage. Regular inspections of plants will allow you to control the pests in the early stages simply by squashing them between your fingers. Severe infestations can be controlled with contact insecticides such as pyrethrum or with pirimicarb which will only kill aphids.

Practical Tips for Spectacular Containers

- Always use fresh potting compost. Choose a soil-based type for permanent plants unless weight is an issue (such as with hanging baskets). Soil-less compost is fine for growing short-lived seasonal plants.

- Thoroughly clean containers that have been used previously. Pests and diseases can survive in debris from former plantings.

- Terracotta and stone containers are porous, so reduce water loss by lining the sides with a layer of newspaper. Soak porous pots in water for about an hour or so before planting.

- Good drainage is vital. Ensure that all containers have drainage holes, put broken crocks or polystyrene in the base, and raise containers a little way off the ground so that water flows away readily.

- Keep containers well watered. Rain is rarely sufficient during the growing season and in hot weather watering may be required twice a day. An automatic watering system is an excellent option for busy gardeners.

- To flourish, plants need food as well as water. Potting compost contains enough nutrients for about six weeks. After that the choice is between regular applications of liquid fertilizer or a one-off dose of controlled-release fertilizer.

- Permanent plants need hardly any maintenance, but remember to top-dress with a little potting compost and controlled-release fertilizer in spring, and to repot the plant when necessary.

Scale insect Small, brown, flattened scales usually appear on leaf undersides and stems, so regular inspections are important to catch this pest in the early stages. A telltale sign is the sticky honeydew exuded by these insects onto the leaves below. Small infestations can be removed with cotton-wool buds dipped in methylated spirits. Larger attacks can be treated with systemic insecticide.

Diseases

Mildew Powdery mildew is common on plants that have suffered from drought. A powdery white coating develops on leaves and stems, and the leaves may turn yellow. Remove and dispose of the worst of the infected leaves and spray the plant with a systemic fungicide. With all fungal diseases, put the infected parts into the dustbin or burn them, as the spores of the disease will persist.

Rust Raised orange spots are usually most common on the undersides of leaves. Remove the leaves that are worst infected and spray with a suitable systemic fungicide. Space plants well apart to encourage plenty of air movement around the foliage.

Viruses A wide range of symptoms include mottled and streaked leaves and distorted growth. There is no cure for viruses, and affected plants should be carefully disposed of. Viruses are transmitted from one plant to another by aphids.

Disorders

Nutrient deficiency Symptoms vary enormously, but yellowing, pale or red-tinged leaves are usually a sign that the plant is in need of fertilizer. Feed immediately with a liquid fertilizer containing trace elements, and then feed during the growing season with either controlled-release fertilizer or regular applications of liquid fertilizer.

Lack of water Plants may wilt in part or completely, leaves and flowers will discolour and die if watering is not done promptly. To re-soak rootballs that have dried out completely, submerge the whole container in a bucket of water for about an hour.

Waterlogging Leaves may yellow and wilt, particularly the older foliage. Rotting of the stems and rootball may occur in severe circumstances. Watch out for waterlogging in plants that are in containers without drainage holes and in containers placed in saucers.

Plant Directory

PLANTS with scented flowers or aromatic foliage come from a number of different plant groups. For ease of reference, in this directory the plants are divided into the following categories: shrubs, climbers, perennials and wild flowers, annuals, biennials, tender perennials, bulbs, herbs, roses and conservatory plants.

Each entry includes advice on cultivation requirements and general information on the plants. The sizes given for the plants decribed are approximate, as this will depend on a number of factors: the size of container that is used, the quality of maintenance (watering, feeding, top-dressing) and the geographical location of your garden (that is, how warm or cold it is).

Delicately scented cowslips (*Primula veris*) coupled with heartsease (*Viola tricolor*) provide a simple yet enchanting spring display.

Shrubs

Choisya ternata
Mexican orange blossom

Flowering time: spring to summer
Height and spread: 1.2 × 0.9m (4 × 3ft)
Position: sun; shelter in cold areas
Propagation: half-ripe cuttings in summer
Pruning: early spring if necessary
Glossy, lobed, evergreen leaves, which are aromatic when bruised, and clusters of scented white flowers. *C.* 'Aztec Pearl' has green, deeply cut leaves; those of *C. ternata* 'Sundance' are golden-yellow.

Daphne odora

Flowering time: late winter to spring
Height and spread: 0.9m (3ft)
Position: sun and shelter; move under cover for winter in all but mild areas
Propagation: layering or half-ripe cuttings in mid-summer
Pruning: none required
A dome-shaped evergreen bearing many small clusters of long-lasting, very fragrant flowers. *D. o.* 'Aureomarginata' has white-edged green leaves and white, pink-flushed flowers. *D. o.* f. *alba* (often known as *D. o.* var. *leucantha*) has green leaves and white flowers. All parts are toxic if eaten, and contact with the sap may irritate the skin.

Lavandula
Lavender

Flowering time: summer
Height and spread: to 45 × 30cm (18 × 12in)
Position: sun
Propagation: half-ripe cuttings in summer
Pruning: trim with shears after flowering and again in spring
A popular evergreen with narrow, grey-green, highly aromatic leaves and fragrant flowers. *L. stoechas* subsp. *pedunculata* (usually sold as *L. s.* 'Papillon') has tubular, blue-purple flowers topped with a pair of winged bracts; it needs winter protection in cold areas. *L. angustifolia* is available in a variety of colours including shades of blue, and pink and white.

Lonicera fragrantissima; *L.* × *purpusii*
Winter-flowering honeysuckle

Flowering time: winter
Height and spread: to 1.2m (4ft)
Position: sun or partial shade
Propagation: half-ripe cuttings in summer
Pruning: after flowering if required
Rounded deciduous shrubs that bear many small, white to cream, strongly scented flowers along the naked stems for a long period in winter.

Mahonia aquifolium 'Smaragd'
Oregon grape

Flowering time: winter to early spring
Height and spread: 0.6 × 0.9m (2 × 3ft)
Position: sun or shade
Propagation: half-ripe cuttings in late summer to autumn
Pruning: after flowering if necessary
A compact evergreen shrub with glossy, pinnate, dark green leaves and clusters of bright yellow flowers.

Myrtus communis
Common myrtle

Flowering time: midsummer to autumn
Height and spread: 0.9–1.2 × 0.6m (3–4 × 2ft)
Position: sun and shelter; move under cover for winter in all but mild areas
Propagation: half-ripe cuttings in late summer
Pruning: spring if necessary
An upright evergreen shrub with small, aromatic, glossy, dark green leaves. White flowers with a conspicuous tuft of stamens are borne over a long period. The leaves of *M. c.* 'Variegata' are edged with creamy-white. Can be trained against a wall or trellis.

Philadelphus
Mock orange

Flowering time: early summer
Height and spread: 0.9m (3ft)
Position: sun or partial shade
Propagation: softwood cuttings in summer
Pruning: cut back flowered shoots after flowering
A compact deciduous shrub bearing masses of white, very sweetly scented flowers. Those of *P.* 'Manteau d'Hermine' are double; those of *P. microphyllus* are single.

Pieris japonica

Flowering time: spring
Height and spread: 0.6–0.9m (2–3ft)
Position: sun or light shade, sheltered from wind in cold areas
Propagation: half-ripe cuttings in mid- to late summer
Pruning: none required
An evergreen shrub with attractive foliage and clusters of urn-shaped, honey-scented flowers. Requires ericaceous (lime-free) compost. The following varieties are compact and suitable for containers. *P. j.* 'Debutante' and *P. j.* 'Purity' have green leaves and white flowers. The flowers of *P. j.* 'Dorothy Wyckoff' are red in bud, opening pink, with foliage that is dark green to bronze. *P. j.* 'Little Heath' and *P. j.* 'Variegata' have green and white leaves and white flowers. The leaves can be toxic if eaten.

Santolina rosmarinifolia subsp. *rosmarinifolia* 'Primrose Gem'

Flowering time: summer
Height and spread: 45cm (18in)
Position: sun
Propagation: half-ripe cuttings in late summer
Pruning: trim after flowering and again in spring
An evergreen shrub with thread-like green leaves, aromatic when crushed, and button-like, pale yellow flowers.

Sarcococca
Christmas box, sweet box

Flowering time: winter
Height and spread: 0.6–0.9m (2–3ft)
Position: sun or shade
Propagation: half-ripe cuttings in late summer
Pruning: thin out overgrown plants in early spring
A neat, upright evergreen with pointed, glossy, dark green leaves and numerous very small tufts of white, strongly scented flowers.

Skimmia japonica

Flowering time: late winter to early spring.
Height and spread: 45–60cm (18–24in)
Position: partial or full shade
Propagation: half-ripe cuttings in late summer
Pruning: none required
A neat, compact, evergreen with rounded, dark green leaves. Showy clusters of flower buds are borne through autumn and winter, opening to white flowers. *S. j.* subsp. *reevesiana* bears red berries in autumn and winter. *S. j.* 'Rubella' has red-edged leaves. The berries are mildly toxic if eaten.

Syringa meyeri var. *spontanea* 'Palibin' (syn. *S. palibiniana*)
Lilac

Flowering time: spring
Height and spread: 0.9–1.2m (3–4ft)
Position: sun
Propagation: layering in early summer
Pruning: thin out after flowering if overgrown
A compact variety of lilac bearing dense panicles of lavender-pink, sweetly scented flowers. Deciduous.

Viburnum × bodnantense

Flowering time: winter
Height and spread: 1.5 × 0.9m (5 × 3ft)
Position: sun or partial shade
Propagation: softwood cuttings in summer
Pruning: after flowering if necessary
An upright, deciduous shrub with clusters of fragrant, pink or white flowers. Eventually grows large but can be kept in a container for several years.

Viburnum × juddii

Flowering time: late spring
Height and spread: 0.9–1.2m (3–4ft)
Position: sun or partial shade
Propagation: softwood cuttings in summer
Pruning: after flowering if necessary
A bushy, deciduous shrub with large heads of white flowers which are pink in bud. The flowers have a strong clove scent.

Climbers

The height and spread of climbers depends on the size of their support. All those below grow to a minimum of 1.8m (6ft).

There are also two climbers listed under Annuals, *Ipomoea alba* and *Lathyrus odoratus* (page 112).

Akebia quinata
Chocolate vine

Flowering time: spring
Position: sun or partial shade
Propagation: half-ripe cuttings in summer or layering in winter
Pruning: immediately after flowering
A twining, deciduous climber with red-purple, chocolate-scented flowers and attractive, five-lobed leaves.

Clematis

Flowering time: late summer to autumn
Position: sun or partial shade
Propagation: seed sown in autumn or layering in summer
Pruning: hard prune in spring
Two vigorous species have scented flowers: *C. flammula* has white, star-shaped, strongly scented blooms; *C. rehderiana* has tubular, yellow and cowslip-scented blooms.

Jasminum officinale
Summer jasmine

Flowering time: summer to autumn
Position: sun or partial shade
Propagation: half-ripe cuttings in summer or layering in winter
Pruning: thin overcrowded plants in late winter or immediately after flowering
A vigorous, twining climber with mid-green pinnate leaves and clusters of white, sweetly scented flowers. The leaves of *J. o.* 'Fiona Sunrise' (syn. *J. o.* 'Frojas') are suffused with gold.

Lonicera
Honeysuckle

Flowering time: summer
Position: sun or partial shade, with roots in the shade
Propagation: half-ripe cuttings in summer
Pruning: cut back to strong new growths after flowering
A popular and vigorous twining climber with strongly scented clusters of flowers. *L. caprifolium* is creamy-white flushed with pink. *L. × italica* (also known as *L. × americana*) is yellow flushed with red-purple. *L. periclymenum* (common honeysuckle) varieties are white and yellow, sometimes streaked with red.

Trachelospermum

Flowering time: summer to autumn
Position: sun and shelter
Propagation: half-ripe cuttings in summer or layering in autumn
Pruning: if overgrown, trim in early spring
A handsome, evergreen, twining climber with very fragrant, jasmine-like flowers. The flowers of *T. asiaticum* are creamy white and yellow-centred, ageing to yellow; those of the less hardy *T. jasminoides* (star jasmine) are pure white.

Wisteria floribunda
Japanese wisteria

Flowering time: early summer
Position: sun
Propagation: grafting (best done by professionals)
Pruning: in midsummer cut back side shoots to 5–6 buds from the main stem; in midwinter shorten further to 2–3 buds
A vigorous, twining, deciduous climber bearing long racemes of fragrant flowers. Many varieties are available in colours including blue, purple, pink and white.

Herbaceous Perennials and Wildflowers

Convallaria majalis
Lily-of-the-valley

Flowering time: spring
Height and spread: 23 × 30cm (9 × 12in)
Position: partial or full shade
Propagation: division in autumn
A low-growing perennial bearing arching stems of small, bell-shaped, strongly scented white flowers among broad green leaves. Seeds may be mildly toxic if eaten.

Dianthus
Pinks

Flowering time: summer
Height and spread: 25–45 × 30cm (10–18 × 12in)
Position: sun
Propagation: cuttings in summer
Old-fashioned and modern pinks bear fragrant flowers in a huge range of varieties and colours above cushions of narrow, green or grey leaves. Some are bicolours while others are prettily laced in a contrasting colour. Old varieties bloom for several weeks only, while modern ones produce several flushes of bloom.

Glechoma hederacea 'Variegata'
Variegated ground ivy

Flowering time: summer
Height and spread: 15 × 90cm (6 × 36in)
Position: sun or partial shade
Propagation: division in spring or autumn
A trailing perennial with rounded, scalloped-edged leaves, pale green margined with white. Grown for its leaves which are aromatic when crushed, although clusters of lilac flowers are produced. Good for growing around the edges of containers.

Iris graminea
Plum-tart iris

Flowering time: summer
Height and spread: 45cm (18in)
Position: sun
Propagation: division in late summer to early autumn
Large, purple-violet flowers which have a strong fruity fragrance are borne among grass-like, bright green leaves. Can be toxic if eaten, and contact with sap can irritate the skin.

Nepeta × faassenii
Catmint

Flowering time: summer
Height and spread: 45–60cm (18–24in)
Position: sun
Propagation: division in autumn or spring
Forms a lax clump of stems clothed with toothed, grey-green leaves, aromatic when crushed, and topped with lavender-blue flowers. Irresistible to cats!

Primula

Flowering time: spring
Height and spread: 20–25cm (8–10in)
Position: sun or partial shade
Propagation: seed sown when ripe or division in autumn or early spring
P. veris (cowslip) bears heads of small, deep yellow, fragrant flowers on short stems. *P. vulgaris* (primrose) bears creamy-yellow, fragrant flowers.

Viola odorata
Sweet violet

Flowering time: spring
Height and spread: 20 × 30cm (8 × 12in)
Position: sun or shade
Propagation: division in spring or autumn
A hardy, spreading plant with rounded green leaves, bearing numerous small, blue or white scented flowers on short stems.

Annuals

Asperula orientalis
Blue woodruff

Flowering time: summer
Height and spread: 30 × 10cm (12 × 4in)
Position: partial or full shade
Propagation: seed
A hardy annual bearing clusters of small, sweetly scented, bright blue 'powder puff' flowers.

Centaurea moschata
(now correctly called *Amberboa moschata*)
Sweet sultan

Flowering time: summer
Height and spread: 60 × 23cm (24 × 9in)
Position: sun
Propagation: seed
A fast-growing hardy annual with upright, slender, branched stems bearing sweetly scented, large, fringed flowers in white, yellow, pink or purple. 'Dairy Maid' has yellow flowers and a chocolate scent. Benefits from staking. Sow directly into the ground or in pots as this plant dislikes root disturbance.

Ipomoea alba
Moonflower

Flowering time: summer
Height and spread: to 2.1m (7ft) high, width according to size of support
Position: sun
Propagation: seed
A twining perennial climber, usually grown as a half-hardy annual. Bears very large, trumpet-shaped white flowers that open in the evening to give off a sweet scent. Does best in a warm, sheltered site.

Lathyrus odoratus
Sweet pea

Flowering time: summer
Height and spread: to 1.8m (6ft) high
Position: sun
Propagation: seed
A hardy annual, this popular climbing plant scrambles up by means of tendrils. Stems of wavy-edged, sweetly scented flowers are produced in a wide range of colours (see page 49 for varieties). Deadhead as soon as blooms fade.

Lobularia maritima
Sweet alyssum

Flowering time: summer
Height and spread: 15 × 30cm (6 × 12in)
Position: sun
Propagation: seed
A low-growing, branching, hardy annual, which bears clusters of scented flowers in white, pink or purple. Clip over after the first flush of bloom to encourage further flowering.

Malcolmia maritima
Virginian stock

Flowering time: summer to autumn
Height and spread: to 40 × 15cm (16 × 6in)
Position: sun but ideally with shade around midday
Propagation: seed
A spreading hardy annual bearing spikes of many sweetly scented flowers in white, pink, red or purple. Compacta Series has a neater habit than the species.

Matthiola
Stock

Flowering time: spring to autumn depending on variety and sowing time
Height and spread: to 60 × 15cm (24 × 6in)
Position: sun
Propagation: seed
A varied group of hardy annuals bearing very sweetly scented flowers (see also under Biennials, page 111). *M. longipetala* (syn. *M. bicornis*; night-scented stock) is a low-growing, sprawling plant with tiny, rose-pink flowers that are powerfully scented at night. 'Starlight Scentsation' is a particularly good form. *M. incana* varieties produce showy, columnar flowers and include Cinderella Series, Midget Series and Ten-week Mixed. See the panel on page 32 for more details.

Nicotiana alata
Tobacco plant

Flowering time: summer
Height and spread: 90 × 30cm (36 × 12in)
Position: sun
Propagation: seed
A half-hardy annual bearing tall stems topped with a number of white flowers, tubular opening to a funnel shape, which are strongly fragrant in the evening. *N.* 'Fragrant Cloud' is a particularly good scented variety. Note: there are many coloured varieties of tobacco plant that are not scented. Contact with the foliage may irritate the skin.

Reseda odorata
Mignonette

Flowering time: summer to autumn
Height and spread: to 60 × 23cm (24 × 9in)
Position: sun or partial shade
Propagation: seed
A hardy annual with branching stems that bear conical heads of tiny, star-shaped, yellowish-green to white, very fragrant flowers. Attractive to bees.

Zaluzianskya capensis
Night phlox

Flowering time: summer
Height and spread: to 45cm (18in)
Position: sun
Propagation: seed
A hardy annual bearing white, sometimes maroon-tinted flowers that are heavily scented, particularly in the evening.

Biennials

Dianthus barbatus
Sweet william

Flowering time: spring to summer
Height and spread: 15–60cm (6–24in)
Position: sun
Propagation: seed
A popular biennial bearing flat heads of many small, sweetly scented flowers in purple-red, pink or white, sometimes bicoloured. A wide range of varieties is available with both single and double flowers. *D. b.* 'Messenger Mixed' flowers around two weeks earlier than other varieties.

Erysimum cheiri (formerly *Cheiranthus cheiri*)
Wallflower

Flowering time: spring
Height and spread: 25–80 × 30cm (10–32 × 12in)
Position: sun
Propagation: seed
An evergreen perennial, usually grown as a biennial, that bears many sweetly scented flowers in short clusters. All the following are compact, to 38cm (15in). Bedder Series has flowers in yellow, orange and red. 'Cloth of Gold' is golden-yellow. 'Harlequin' offers a wide range of colours.

Hesperis matronalis
Sweet rocket, dame's violet

Flowering time: spring to summer
Height and spread: 90 × 30cm (36 × 12in)
Position: sun or partial shade
Propagation: seed
This tall biennial produces stout stems topped with heads of lilac, purple or white flowers that are strongly scented in the evening. Usually needs staking.

Matthiola incana
Stock

Flowering time: summer
Height and spread: to 60 × 30cm (24 × 12in)
Position: sun
Propagation: seed
Certain strains of stock, such as Brompton Series, East Lothian Series and Legacy Series, can be grown as biennials. They have scented flowers in a range of colours, including red, pink, lavender-blue and creamy-yellow. See panel on page 32 for more details.

Tender Perennials

Cosmos atrosanguineus
Chocolate cosmos

Flowering time: summer
Height and spread: 60 × 45cm (24 × 18in)
Position: sun
Propagation: basal cuttings in early spring
An upright to spreading perennial bearing single, cup-shaped, dark red flowers that smell of chocolate. Deadhead to prolong flowering. In cold areas, lift tubers before the first frosts and overwinter in trays of barely moist compost in a frost-free place.

Heliotropium
Heliotrope, cherry pie

Flowering time: summer
Height and spread: 45cm (18in)
Position: sun
Propagation: half-ripe cuttings in summer
A bushy, short-lived shrub, which is often grown as an annual. Numerous tiny flowers, borne in large, dense flowerheads, are attractive to bees. See the project on page 50 for varieties. Contact with foliage may irritate the skin.

Mirabilis jalapa
Marvel of Peru, four o'clock plant

Flowering time: summer
Height and spread: 60cm (24in)
Position: sun
Propagation: seed
A bushy, tuberous perennial that bears many small, scented flowers in red, pink, yellow or white, which open in late afternoon. In cold areas, lift tubers in autumn and store in trays of barely moist compost in a frost-free place for the winter.

Nemesia denticulata (syn. *N. d.* 'Confetti') and *N.* 'Fragrant Cloud'

Flowering time: summer
Height and spread: 45cm (18in)
Position: sun
Propagation: cuttings of shoot tips in late summer
These lax, slightly spreading perennials bear a profusion of small clusters of pale pink, scented flowers on slender stems. Will tolerate light frosts.

Pelargonium
Scented-leaved types

Flowering time: summer
Height and spread: to 60cm (24in)
Position: sun
Propagation: cuttings in spring or in late summer to autumn
A delightful evergreen perennial with attractively shaped, sometimes variegated leaves that give off a strong and distinct perfume when crushed. Small, single flowers are mauve, pink, purple or white, but these are of secondary interest to the foliage. See the panel on page 51 for varieties.

Bulbs

Crocus chrysanthus

Flowering time: late winter
Height and spread: 5cm (2in)
Position: sun
Propagation: seed sown as soon as ripe
This early-flowering bulb bears up to four scented flowers on short stems. Add grit to the potting compost to ensure good drainage. *C. c.* 'Cream Beauty' is creamy-yellow; *C. c.* 'E.A. Bowles' is golden-yellow; *C. c.* 'Snow Bunting' is white.

Freesia

Flowering time: winter to spring (prepared corms bloom in summer)
Height and spread: to 40 × 8cm (16 × 3in)
Position: sun
Propagation: remove offsets in autumn
These popular cut flowers are extremely sweetly scented and are available in a range of colours, including yellow, blue and white. Plant prepared corms under cover in spring to bloom outdoors in summer. Alternatively, plant under cover in autumn and keep protected for spring flowering.

Galanthus
Snowdrop

Flowering time: late winter
Height and spread: to 20 × 8cm (10 × 3in)
Position: partial shade
Propagation: division of established clumps immediately after flowering
This popular bulb bears honey-scented pure white flowers, delicately marked with green inside. *G. nivalis* 'Flore Pleno' has double flowers; *G.* 'S. Arnott' bears large, strongly scented flowers. Mildly toxic if eaten; contact with bulbs may irritate the skin.

Hyacinthus orientalis
Hyacinth

Flowering time: spring
Height and spread: 20–30 × 8cm (8–12 × 3in)
Position: sun or partial shade
Propagation: remove offsets in summer
This popular bulb bears large, showy heads of flowers that are very strongly scented and are available in many colours, including blue, purple, yellow, pink and white. Bulbs growing outside are susceptible to frost damage, so protect them if necessary. Specially prepared bulbs can be planted in late summer to early autumn to flower indoors during winter. Toxic if eaten; contact with bulbs may irritate sensitive skin.

Iris reticulata

Flowering time: late winter
Height and spread: 10–15 × 5cm (4–6 × 2in)
Position: sun
Propagation: separate overcrowded clumps in late autumn
A dwarf iris bearing small, violet-blue flowers, the lower petals marked with yellow.

Lilium

Flowering time: mid- or late summer
Height and spread: 30–120 × 10cm (12–48 × 4in)
Position: sun
Propagation: seed (of species) or removal of bulblets when foliage dies back
Large, cup- to bowl- or trumpet-shaped flowers are borne in summer. Those listed below have a rich perfume, although there are a number of lilies that have little or no scent.

Species
L. auratum (golden-rayed lily of Japan, mountain lily) has pure white flowers with a bold yellow band down the centre of each petal in late summer. It is a gorgeous lily but not the easiest to grow, being prone to virus and preferring a sheltered spot in light, dappled shade; 1.2–1.5m (4–5ft) high. *L. formosanum* var. *pricei* produces pure white, trumpet-shaped blooms on short, stout stems in late summer; 30–60cm (12–24in) high. *L. longiflorum* (Easter lily, Bermuda lily) blooms in midsummer, bearing pure white, trumpet-shaped flowers; 60cm (24in) high. *L. regale* (regal lily) blooms in midsummer and is the easiest lily of all to grow, with a superb perfume. Trumpet-shaped white flowers have a yellow throat and are flushed with purple outside. *L. r.* 'Album' is pure white with a yellow throat; 0.9–1.2m (3–4ft).

Hybrids
Oriental hybrids bloom in mid- to late summer. They prefer to have their heads in the sun and their roots in the shade, and shouldn't be allowed to dry out. All grow to 90–120cm (3–4ft) unless stated otherwise. *L.* 'Casa Blanca' has exceptionally large blooms, which are pure white, with orange-brown stamens. *L.* 'Hit Parade' is rose-pink fading to white in the centre, with petals slightly curved back at the tips. *L.* Imperial Gold Group has white flowers banded with gold and spotted with crimson. *L.* Imperial Silver Group has very large white flowers, spotted with red. *L.* 'Journey's End' has rich crimson-pink flowers edged with white. *L.* 'Kiss Proof' has deep crimson-red flowers with a narrow white margin. *L.* 'Mona Lisa' has soft pink flowers; 40cm (16in) high. *L.* 'Noblesse' has pale rose-pink flowers speckled with brown, with a yellow band down the centre of each petal. *L.* 'Star Gazer' has deep crimson flowers with a white edge; 60cm (24in) high.

Trumpet lilies
Trumpet lilies bloom in midsummer and prefer full sun; they grow to 1.2m (4ft) high. *L.* African Queen Group has large apricot flowers. *L.* Golden Splendour Group has deep golden-yellow flowers. *L.* Green Magic Group has lemon-yellow flowers shaded with green. *L.* Pink Perfection Group has large, deep pink flowers.

Muscari armeniacum
Grape hyacinth

Flowering time: spring.
Height and spread: 20 × 8cm (8 × 3in)
Position: sun or partial shade
Propagation: seed
The conical heads are made up of many small, bright blue flowers, with a honey scent, which are produced on short stems above grassy foliage. The leaves appear in autumn.

Narcissus
Daffodil

Flowering time: spring
Height and spread: 15–45 × 8cm (6–18 × 3in)
Position: sun
Propagation: separate established clumps in early autumn
These popular bulbs bear flowers in shades of yellow and white, sometimes bicoloured. Although all narcissi are scented to some degree, the three groups detailed below contain the most fragrant varieties. Contact with sap may irritate the skin.

Doubles
Flowering in early to mid-spring, the stems bear one or more double blooms. Grow in sun or light shade.
N. 'Cheerfulness', creamy-white; 40cm (16in). *N.* 'Flower Drift', white outer petals, red centre; 35cm (14in).
N. 'Pencrebar', golden-yellow; 18cm (7in). *N.* 'Sir Winston Churchill', white with orange florets; an outstanding variety; 40cm (16in). *N.* 'Yellow Cheerfulness', golden-yellow; 45cm (18in).

Jonquilla
Flowering in mid- to late spring, each stem bears one to five flowers with small, shallow cups. Require full sun all year.
N. 'Baby Moon' golden-yellow; 20cm (8in). *N. jonquilla* single, lemon-yellow; 30cm (12in). *N.* 'Pipit' soft lemon, the cups ageing to white; 23cm (9in). *N.* 'Quail' golden-yellow with a profusion of flowers; 20cm (8in). *N.* 'Suzy' golden-yellow petals and orange cups; 50cm (20in). *N.* 'Trevithian' lemon-yellow; 45cm (18in).

Tazetta
Flowering in early to mid-spring, each stem bears from three to 20 small blooms. Require full sun all year.
N. 'Geranium' white with an orange cup; 35cm (14in).
N. 'Grand Soleil d'Or' golden-yellow with an orange cup; 45cm (18in). *N.* 'Minnow' creamy-white with a lemon cup; 20cm (8in). *N.* 'Silver Chimes' creamy-white with a primrose-yellow cup; 30cm (12in).

Tulipa
Tulip

Flowering time: spring
Height and spread: to 38 × 10cm (15 × 4in)
Position: sun
Propagation: separate and replant offsets in summer
Several tulips have a sweet scent, although it is nowhere near as strong as that of many other bulbs. *T. tarda* bears star-shaped flowers that are bright yellow tipped with white and grows to 15cm (6in) high. Hybrids include: *T.* 'Bellona', golden-yellow; *T.* 'Black Parrot' deep purple; *T.* 'Generaal de Wet' golden-orange; *T.* 'Orange Favourite', orange marked with green and with a yellow base.

Herbs

Allium schoenoprasum
Chives and
A. tuberosum
Garlic chives

Flowering time: summer
Height and spread: 30–40 × 15cm (12–16 × 6in)
Position: sun
Propagation: division in early spring or seed in late spring
These useful perennial culinary herbs have narrow, grass-like, onion- or garlic-flavoured leaves and purple or white flowers. They need plenty of water to stay green and fresh in containers.

Aloysia triphylla (syn. *Lippia citriodora*)
Lemon verbena

Flowering time: summer
Height and spread: 1.2–1.5m (4–5ft)
Position: sun and shelter
Propagation: half-ripe cuttings in summer to autumn

A slender-stemmed shrub with long, slender, light green leaves that smell strongly of lemon when bruised, and panicles of lilac-coloured flowers. Can be grown as a bushy shrub, trained as a standard or trained closely against a wall or trellis. Move under cover in winter in cold areas.

Anethum graveolens
Dill

Flowering time: summer
Height and spread: 60–90 × 30cm (24–36 × 12in)
Position: sun
Propagation: seed
A hardy annual with fine, feathery, aromatic green foliage and heads of yellow flowers. Needs staking. Cut regularly to keep plants compact and to encourage young growth.

Calamintha grandiflora 'Variegata'
Calamint

Flowering time: summer
Height and spread: 30cm (12in)
Position: sun
Propagation: cuttings of young shoots in spring
A perennial with oval, toothed leaves that smell strongly of mint when crushed, and small clusters of lilac flowers. *C. g.* 'Variegata' has attractive cream and green leaves.

Foeniculum vulgare 'Purpureum'
Bronze fennel

Flowering time: summer
Height and spread: to 1.5m × 45cm (to 5ft × 18in)
Position: sun
Propagation: seed
This tall, architectural perennial has feathery, bronze-green foliage that smells of aniseed when crushed, and flat heads of yellow flowers. It selfseeds freely. Shelter from the mid-day sun in summer.

Galium odoratum
Sweet woodruff

Flowering time: spring to summer
Height and spread: 15 × 30cm (6 × 12in)
Position: partial or full shade
Propagation: seed in autumn or root cuttings in early summer
A shade-loving perennial with divided green leaves and white star-shaped flowers, all of which have the scent of new-mown hay that is strongest when dried. Excellent for under-planting shrubs.

Helichrysum italicum subsp. *serotinum*
Curry plant

Flowering time: summer
Height and spread: 45–60cm (18–24in)
Position: sun
Propagation: half-ripe cuttings in summer to autumn
A shrubby herb with narrow, attractive, silvery leaves that smell strongly of curry, particularly after rain. Heads of bright yellow flowers are also produced. Trim after flowering and again in spring. Take care not to overwater.

Laurus nobilis
Sweet bay

Flowering time: spring
Height and spread: as trained, usually to 1.8m (6ft)
Position: sun
Propagation: layering in spring
An evergreen shrub with aromatic dark green leaves that are widely used in cooking. Can be clipped to a variety of shapes, such as cones and lollipops. Move under cover for winter in all but mild areas, as the leaves can be scorched by hard frosts. Do not over-water at any time, and keep on the dry side in winter.

Melissa officinalis 'Aurea'
Lemon balm

Flowering time: summer
Height and spread: 60 × 30cm (24 × 12in)
Position: sun
Propagation: division in autumn or spring
A perennial herb with green leaves brightly marked with yellow, and with a strong lemon scent when crushed. Cut back regularly to encourage good variegation. Best moved under cover for winter in cold areas.

Mentha
Mint

Flowering time: summer to early autumn
Height and spread: height depends on species; spread indefinite
Position: sun
Propagation: division in spring or autumn
This popular herb has leaves that are pungently scented when crushed. It is invasive, spreading by underground runners, so do not grow with other plants. In autumn, bring a small potful indoors on a warm windowsill and fresh shoots should grow in a couple of weeks. See panel on page 67 for varieties.

Ocimum basilicum
Basil

Flowering time: summer
Height and spread: 30cm (12in)
Position: sun and shelter
Propagation: seed sown direct because basil dislikes being transplanted
A half-hardy annual with strongly aromatic leaves that are widely used in cooking. Varieties with purple leaves are particularly decorative; they include *O. b.* 'Dark Opal', *O. b.* 'Purple Ruffles' and *O. b.* 'Red Rubin'. Do not overwater, and avoid evening watering if possible so that the plant does not have 'wet feet' overnight.

Origanum
Marjoram

Flowering time: summer
Height and spread: 30–45cm (12–18in)
Position: sun
Propagation: cuttings in spring
A decorative perennial herb with rounded, aromatic leaves and pale pink flowers. *O. vulgare* 'Aureum' (golden marjoram), which has golden leaves and is ideal for culinary use, benefits from midday shade in summer. *O.* 'Kent Beauty' is decorative rather than useful, with mid-green leaves and showy, funnel-shaped flowers surrounded by conspicuous yellow-green bracts.

Petroselinum crispum
Parsley

Flowering time: summer
Height and spread: 30cm (12in)
Position: partial shade
Propagation: seed
A popular hardy biennial culinary herb. Varieties with curled green leaves are most decorative. Make several sowings throughout the year to ensure a regular supply, because parsley runs to seed quickly in its second year. Prefers midday shade in summer but site in full sun in winter.

Rosmarinus officinalis
Rosemary

Flowering time: spring to summer
Height and spread: 30–120 × 60cm (12–48 × 24in)
Position: sun
Propagation: cuttings in summer
This popular shrub has slender, densely clustered, dark green leaves and masses of small blue flowers. *R. o.* 'Miss Jessopp's Upright' forms a narrow, upright bush, while *R. o.* Prostratus Group (also known as *R. lavandulaceus*) is low and spreading. The latter is more tender and should be moved under cover in cold areas. Feed rosemary only after flowering.

Salvia officinalis
Sage

Flowering time: summer
Height and spread: to 60cm (24in)
Position: sun
Propagation: cuttings in late spring to early summer
Several varieties of this hardy perennial sage have aromatic, coloured or variegated leaves that are very decorative as well as edible. The flowers are blue-mauve. *S. o.* 'Icterina' (golden sage) is green and gold. *S. o.* Purpurascens Group is purple. *S. o.* 'Tricolor' is variegated with green, pink, white and purple. Flowers are less freely produced on these forms than on the green-leaved species. Clip in spring to maintain a bushy plant. Do not overwater.

Satureja montana
Winter savory

Flowering time: summer
Height and spread: 30 × 20cm (12 × 8in)
Position: sun
Propagation: cuttings in spring
A semi-evergreen hardy perennial that forms a small, neat bush with narrow, very aromatic leaves and bears small, white and pink flowers. Trim regularly to keep plants bushy. Give no more than one or two liquid feeds during the growing season. Move under cover for winter in cold areas. Can be planted as a low, informal hedge in warmer areas, in which case set individual plants 45cm (18in) apart.

Thymus
Thyme

Flowering time: summer
Height and spread: 10 × 20cm (4 × 8in)
Position: sun
Propagation: layering in spring
A popular evergreen hardy perennial that forms a low mound or mat of tiny aromatic leaves and flowers. A wide selection of varieties includes *T.* × *citriodorus* 'Archer's Gold' (lemon thyme), which has lemon-scented green and gold leaves; *T.* 'Doone Valley' has similar leaves; *T. serpyllum* varieties are creeping in habit, with pink, red, white or purple flowers. Water and feed sparingly. Trim after flowering to keep plants bushy.

Roses

Miniature climbers

Flowering time: summer to autumn
Height and spread: 1.8 × 1.2m (6 × 4ft)
Position: sun
Propagation: hardwood cuttings in autumn to winter
A relatively new group of compact climbers, often called patio climbers, that can be grown in a large container such as a half-barrel.
R. 'Little Rambler' (syn. *R.* 'Chewramb') has small, pale pink to white flowers, with a strong scent; the stems are lax and pliant, similar to those of a larger rambler rose.
R. 'Nice Day' (syn. *R.* 'Chewsea') is salmon pink with a sweet scent.
R. 'Warm Welcome' (syn. *R.* 'Chewizz') is orange with a faint fragrance. Prune in spring, cutting back lateral side shoots to 2–3 buds from the main stem. Tie in stems after pruning and every 2–3 weeks during the growing season.

Miniature, dwarf floribunda and ground-cover roses

Flowering time: summer to autumn
Height and spread: see individual varieties
Position: sun
Propagation: hardwood cuttings in autumn to winter
Although these three groups of roses are ideal for containers, few are scented. As their name suggests, miniature roses bear tiny blooms, and they are the smallest of this group. Dwarf floribunda varieties have larger flowers and are a little taller. These two types of rose are often grouped together and called 'patio roses'. Ground-cover roses are spreading in habit. Feed regularly and deadhead to prolong flowering.
R. 'Colibre '79' is a miniature with double yellow flowers flushed with pink; H and S 30cm (12in).
R. 'Dresden Doll' is a miniature with semi-double, shell-pink flowers; H and S 30cm (12in).
R. 'Golden Angel' is a miniature with double, deep yellow flowers; H and S 30cm (12in).
R. 'Norfolk' (syn. *R.* 'Poulfolk), a ground-cover rose, has double, bright yellow, very fragrant flowers over a long period; H 60cm (24in), S 90cm (36in).
R. 'Peachy White', a miniature, has white flowers, flushed pink; H and S 30cm (12in).
R. 'Pink Posy' (syn. *R.* 'Cocanelia') is a dwarf floribunda with double, rosy-pink, sweetly scented flowers borne in sprays; H and S 60cm (24in).

R. 'Regensberg' (syn. *R.* 'Macyou') is a dwarf floribunda with double, bright pink and white flowers; H and S 45cm (18in).
R. 'Sweet Dream' (syn. *R.* 'Fryminicot'), a dwarf floribunda, has slightly fragrant, cupped, apricot-peach flowers, borne on a bushy, upright plant; H and S 45cm (18in).
R. 'Sweet Magic' (syn. *R.* 'Dicmagic') is a dwarf floribunda with orange flowers tinted with gold and a slight fragrance; H and S 45cm (18in).
R. 'Yorkshire', a ground-cover rose, bears semi-double, white flowers through summer and autumn; H 60cm (24in), S 90cm (36in).

Conservatory Plants

All the following plants must have a frost-free environment all year. See pages 88–95 for growing details.

Boronia megastigma

Flowering time: spring
Height and spread: to 1.5 × 0.9m (5 × 3ft)
Position: good light but shaded from hot sun
Propagation: half-ripe cuttings in summer
Pruning: immediately after flowering cut back flowered shoots to within 2.5cm (1in) of the previous year's growth
This upright shrub has slender, aromatic, deep green leaves. Downward-facing, bell-shaped, lemon-scented flowers are reddish-brown outside and yellow-green inside. Needs ericaceous (lime-free) potting compost.

Brugmansia
Angel's trumpets

Flowering time: summer
Height and spread: to 1.8m (6ft)
Position: full light
Propagation: half-ripe cuttings in summer
Pruning: on overgrown plants cut back several stems to 2–3 buds in spring; repeat the following year
Formerly known as *Datura*, this tall shrub bears huge, showy, trumpet-shaped blooms that hang downwards and release a potent scent. A number of species and varieties are available, with colours including yellow, orange, white, pink and apricot. All parts of the plant are very poisonous.

Buddleja asiatica; B. auriculata

Flowering time: winter
Height and spread: to 1.8m (6ft)
Position: good light but shaded from full sun in summer
Propagation: half-ripe cuttings in summer
Pruning: cut back hard in spring
Evergreen shrubs bearing fragrant white flowers. Those of *B. auriculata* have orange or pink centres.

Camellia sasanqua 'Narumigata'

Flowering time: autumn
Height and spread: 1.5 × 0.9m (5 × 3ft)
Position: good light but shaded from sun
Propagation: half-ripe cuttings in late summer to autumn
Pruning: after flowering if necessary
An elegant evergreen shrub with oval, dark green leaves and an upright habit. Bears fragrant, single flowers that are white tinged with pink. Grow in ericaceous (lime-free) potting compost and in hard water areas avoid watering with tap water if possible.

Citrus × meyeri 'Meyer'
Lemon

Flowering time: mostly spring to summer, but blooms are usually produced all year
Height and spread: 1.2 × 0.9m (4 × 3ft)
Position: full light, shaded from sun

Propagation: half-ripe cuttings in summer; seed can be sown in spring but seedlings will not come true
Pruning: in spring, only if necessary
A branching evergreen shrub with aromatic leaves and clusters of strongly scented white flowers. See panel on page 91 for growing details.

Dregea sinensis (syn. *Wattakaka sinensis*)

Flowering time: summer
Height and spread: to fit support
Position: good light or light shade
Propagation: stem cuttings in summer to autumn
Pruning: as required after flowering
A twining evergreen climber with heart-shaped leaves and clusters of creamy-white, pink-streaked, fragrant flowers. Will tolerate light frosts but is best moved under cover for winter in cold areas.

Eriobotrya japonica
Loquat

Flowering time: autumn to winter
Height and spread: 1.8–2.4m (6–8ft)
Position: good light
Propagation: half-ripe cuttings in summer or seed in spring
Pruning: hard prune in spring if necessary to restrict growth
A large evergreen shrub with huge, architectural, dark green leaves that are distinctively ridged. Large panicles of fragrant white flowers are followed by edible, orange-yellow fruit.

Eucalyptus citriodora
Lemon-scented gum

Flowering time: flowers not borne on plants that are kept cut back
Height and spread: to 1.8m × 60–90cm (6ft × 24–36in)
Position: full sun
Propagation: seed in spring or summer
Pruning: cut back hard each year in spring to keep to 1.5–18m (5–6ft) high
A tall- and fast-growing plant in the open ground, lemon-scented gum should be cut back hard each year to keep its height within bounds for a conservatory and for the production of pleasantly scented young foliage.

Euphorbia mellifera (syn. *E. longifolia*)
Honey spurge

Flowering time: late spring
Height and spread: to 1.5m (5ft)
Position: good light
Propagation: cuttings of stem tips in spring to early summer
A rounded evergreen shrub with long, narrow, dark green leaves and clusters of honey-scented flowers. Toxic if eaten; contact with sap may irritate the skin.

Jasminum azoricum
Jasmine

Flowering time: late summer
Height and spread: to fit support
Position: good light with shade from hot sun
Propagation: half-ripe cuttings in summer or layering in autumn
Pruning: thin out growth after flowering if overcrowded
A twining evergreen climber with deep green leaves and fragrant flowers, which are purple in bud opening white.

Jasminum polyanthum

Flowering time: winter to spring
Height and spread: to fit support
Position: good light with shade from hot sun
Propagation: half-ripe cuttings in summer or layering in autumn
Pruning: thin out growth after flowering if overcrowded
A vigorous, twining evergreen climber widely sold as a pot plant. Masses of very strongly scented flowers are pink in bud opening white.

Mandevilla laxa (syn. *M. suaveolens*)
Chilean jasmine

Flowering time: summer to autumn
Height and spread: to fit support

Position: good light with shade from hot sun
Propagation: half-ripe cuttings in summer
Pruning: in late winter to early spring cut back side shoots to 3–4 buds and thin out other growth if necessary
A vigorous, twining climber with rich green leaves and racemes of strongly scented, tubular, white flowers. Mildly toxic if eaten, and contact with sap may irritate the skin.

Pittosporum tobira
Japanese pittosporum, Japanese mock orange

Flowering time: spring to summer
Height and spread: to 1.8m (6ft)
Position: good light, shaded from hot sun
Propagation: half-ripe cuttings in summer or seed sown as soon as ripe
Pruning: hard prune after flowering if necessary to restrict growth
A large, rounded, evergreen shrub with leathery, dark green leaves and large clusters of very sweetly scented, creamy-white flowers. Can be grown as a standard.

Primula kewensis

Flowering time: winter to spring
Height and spread: 45 × 20cm (18 × 8in)
Position: good light with shading from hot sun
Propagation: seed in early spring or division after flowering
An evergreen perennial, bearing stems topped with clusters of fragrant yellow flowers.

Prostanthera cuneata
Alpine mint bush

Flowering time: summer
Height and spread: 45–60cm (18–24in)
Position: good light
Propagation: half-ripe cuttings in summer
An upright to spreading small shrub with rounded leaves that are very aromatic when crushed, and long racemes of tubular white flowers marked purple and yellow within.

Salvia elegans 'Scarlet Pineapple' (syn. *S. rutilans*)
Pineapple sage

Flowering time: summer
Height and spread: 90 × 60cm (36 × 24in)
Position: good light but shaded from hot sun
Propagation: softwood cuttings in spring or half-ripe cuttings in late summer
A tender perennial with bright red flowers and green, red-tinged leaves that have a gorgeous pineapple scent when bruised.

Index

Acknowledgements

Photographs

John Glover 8 (design: Jane Fearnley-Whittingstall), 9 (styling and photography: John Glover), 10 (John Glover), 12 (John Glover), 13 (Sticky Wicket, Dorset), 19, 20 (John Glover), 23 (Jekka McVicar's herbs), 27 (John Glover), 46, 62 (John Glover), 69 (Whichford Pottery, Warkwickshire), 81, 82 (John Glover), 85 (John Glover), 88 (design: Diarmuid Gavin), 91 (design: Bunny Guinness), 97 (design: Susy Smith), 106 (design: Marnie Hall); **Jerry Harpur** 7 (design: Anne Alexander-Sinclair, London), 52, 77 (design: Dan Pearson, London), 78 right (design: Susie Ind, London); **Marcus Harpur** 86; **Anne Hyde** 45 (Mr and Mrs Coote, Oxford), 55 (design: Ben Loftus); 65 (Howard and Audrey Pring, Lower Severalls Herb Garden, Somerset), 67 (Knebworth House, Hertfordshire), 68 (Howard and Audrey Pring, Lower Severalls Herb Garden, Somerset); **Andrew Lawson** 6 , 14, 18 (Bourton House, Gloucestershire), 22, 24, 25, 26, 40, 41 (The Old Rectory, Sudborough, Northamptonshire) 42, 43, 47, 48 right, 48 left, 51, 57, 64 right, 64 left, 72 right, 72 left, 76, 78 left, 79, 90, 93; **Clive Nichols** 16, 29 (design: Graham Strong), 56 (The Old Rectory, Berkshire), 58 (design: Guy Farthing).

Artwork

Wendy Bramall